Composites and Metamaterials

Composites and Metamaterials

Roderic Lakes
University of Wisconsin-Madison, USA

World Scientific
NEW JERSEY • LONDON • SINGAPORE • BEIJING • SHANGHAI • HONG KONG • TAIPEI • CHENNAI • TOKYO

Published by

World Scientific Publishing Co. Pte. Ltd.

5 Toh Tuck Link, Singapore 596224

USA office: 27 Warren Street, Suite 401-402, Hackensack, NJ 07601

UK office: 57 Shelton Street, Covent Garden, London WC2H 9HE

Library of Congress Cataloging-in-Publication Data
Names: Lakes, Roderic S., author.
Title: Composites and metamaterials / Roderic Lakes, University of Wisconsin-Madison, USA.
Description: New Jersey : World Scientific, [2020] |
Includes bibliographical references and index.
Identifiers: LCCN 2019057935 | ISBN 9789811216367 (hardcover) | ISBN 9789811216374 (ebook) |
ISBN 9789811216381 (ebook other)
Subjects: LCSH: Composite materials--Mechanical properties. |
Metamaterials--Mechanical properties.
Classification: LCC TA418.9.C6 L35 2020 | DDC 620.1/18--dc23
LC record available at https://lccn.loc.gov/2019057935

British Library Cataloguing-in-Publication Data
A catalogue record for this book is available from the British Library.

For any available supplementary material, please visit
https://www.worldscientific.com/worldscibooks/10.1142/11715#t=suppl

Desk Editors: Tay Yu Shan/Nur Izdihar Binte Ismail

Preface

This book is intended to be of use in a one semester course on the structure and properties of materials that are heterogeneous on size scales larger than the atomic and the nano. Emphasis is placed on mechanical properties; also the coupling of elastic behavior with electric and thermal variables. Draft manuscript materials have been used for an interdisciplinary course for senior undergraduate students and graduate students in materials science, mechanics, biomedical engineering, and mechanical engineering. The assumed prerequisite subjects include introductory university physics and a course in the mechanical properties of materials. Sections on novel classes of materials with unusual or extreme behavior are intended both for students and for researchers.

As for materials, both well-established materials and also novel materials that have been popular in a research setting are presented. For established materials, selected applications are presented. The intent is not to be encyclopedic. Lamination theory, for example, is a substantial part of many composites courses and is thoroughly covered elsewhere. In the present book, this subject is summarized briefly. To help clarify the flow of ideas, primary references are used when they are known, and the evolution of the thought process is outlined.

Solved problems are provided for instructional use; also problems without solutions. Supplementary material, resources for instruction and research links are provided in the following web site.

`http://silver.neep.wisc.edu/~lakes`

I thank colleagues in the Department of Engineering Physics and Department of Materials Science at the University of Wisconsin–Madison for encouragement and support during a portion of the time this book was in preparation. I am also grateful to my wife Diana for her patience and support.

Contents

Chapter 1

Introduction

1.1 Heterogeneous materials

1.1.1 Overview

Composite materials are *heterogeneous* materials which contain two or more distinct constituent materials or phases, on a microscopic or macroscopic size scale larger than the atomic scale, and in which physical properties are significantly altered in comparison with those of a homogeneous material. In this vein, fiberglass and other fibrous materials are viewed as composites, but alloys such as brass are not. Semi-crystalline polymers such as polyethylene, as well as eutectics, polycrystalline metals, have a heterogeneous structure which can be treated via composite theory. Biological materials such as bone, tendon and muscle, and geological materials such as rocks and minerals also have a heterogeneous structure and are known as natural composites. Composites may contain solid, liquid, and gas phases. For example, composites of gas and liquid include mist and foam; composites of solid and gas include foam and smoke. Composites with a structural role have several solid phases or a connected solid phase with gas or liquid in the interstices (as in cellular solids such as structural foam and honeycomb). *Lattices* comprise a class of cellular solids in which there is a repeated pattern of cells. Many composites are *anisotropic*: their properties depend on direction. Anisotropy of properties arises from structural anisotropy. Fibrous composites with fully or partially aligned fibers can be highly anisotropic. Fibers used in composites are much stronger than bulk materials because they can be prepared with few defects. Composites are of particular interest in applications for which *density* [1] is important in relation to physical properties.

Materials may have a variety of physical properties that are coupled. Such coupled fields are of scientific interest; they are also the basis of many practical applications including sensors, actuators and devices to extract energy from the environment (energy harvesting).

Analytical study of composite materials has been done for more than 100 years [2]. For example, James Clerk Maxwell [3] in 1873 reported an analysis of the conductivity of an array of conducting spheres in a conducting matrix.

Composite materials are useful because they may combine beneficial characteristics of the constituent materials or they may provide characteristics not present in the constituents. For example, wood is stiff and strong in one direction only. To improve properties in other directions, plywood is made: the wood is cut in layers which are cemented together in different directions. The resulting plywood is relatively stiff and strong in two directions along the plane of the board. Also, it is possible in composites to achieve physical properties greater than that of any constituent. In heterogeneous materials one can also achieve unusual behavior such as negative physical properties, or extreme behavior.

1.1.2 Classification and terminology

Some terminology and neologisms have emerged regarding heterogeneous materials. *Smart materials* are materials designed such that one or more properties can be altered or controlled by field variable stimuli. These stimuli include but are not limited to stress, temperature, humidity, fluid pressure, pH, electric field, and magnetic field. *Bio-mimetic materials* are materials in which the design imitates structures observed in biological materials. *Bio-inspired materials* are materials in which the design process is influenced by but is not directly imitative of structures observed in biological materials. Materials that admit coupled fields are pertinent in that regard. *Auxetic materials* have a negative Poisson's ratio. *Metamaterials* are materials in which structure and composition allows unusual properties such as negative values of physical properties, or extreme values of physical properties; meta means beyond or superior. Mechanical or elastic metamaterials provide extreme elastic properties. Electromagnetic metamaterials allow control of electromagnetic waves. Use of 3D printing, while certainly useful, is not a necessary requirement for metamaterials. Moreover, making a material with microstructure via 3D printing does necessarily confer extreme or unusual properties. If not, calling such a material a metamaterial is inconsistent with the definition. *Architected materials* are multiphase materials or cellular materials that are controlled and optimized to achieve specific functions or properties. *Multifunctional* materials are those that serve multiple roles in a practical application.

The terminology is recent but representative materials in each category are hardly new. Multifunctional materials have been with us for centuries. For example, sod and mud brick used to make dwelling places for hundreds or thousands of years will serve multiple functions such as a structural role in protecting people from rain and wind as well as to provide thermal insulation insulating people from extremes in the temperature of the environment. Smart materials make use of coupled fields. Piezoelectricity and pyroelectricity, for example, have long been known. The permeability of rock, soil, building material and other materials has been known for millennia. The first negative Poisson's ratio foams date back more than 30 years and predate the words auxetic and metamaterial. As for architected materials, all composites including the earliest ones may be considered to be within that category. The term is usually used in the context of composite or cellular solid structures that are facilitated by additive manufacturing techniques such as 3D printing. In this book, the actual properties of the materials are emphasized rather than neologisms; the recent terminology may be found in the index.

1.1.3 Assumptions about the material

Introductory treatments of material properties generally assume the material is homogeneous, isotropic, linear and elastic. These are independent assumptions. The first assumption, that of homogeneity, is relaxed in all composite materials. If the size of the largest heterogeneity is sufficiently small, the usual classical continuum assumptions are usually warranted. In a continuum, one uses concepts, such as stress and strain, that are defined in the sense of passage to a limit of a differential element. If the heterogeneity size is not negligible, nonclassical effects occur as discussed in Chapter 8. Anisotropy refers to the direction dependence of properties as discussed in Chapter 3. A material may be macroscopically homogeneous, for example a single crystal, and be anisotropic. By contrast a material may be heterogeneous, for example a material containing particles or fibers that are randomly distributed and oriented, and be isotropic. As for linearity, most materials, even those that are heterogeneous and/or anisotropic, are linear for an input cause that is sufficiently small. Nonlinearity or fracture can occur for a sufficiently large input cause. Inelastic behavior for deformable materials refers to a difference between loading and unloading. Such a difference could be a result of yield or fracture under sufficiently large stress. The path for unloading may differ from the path for loading if the material is viscoelastic (has properties that depend on time or frequency). Viscoelastic behavior may be linear or nonlinear as discussed in Chapter 9.

Assumptions about the material may not always be stated explicitly. Such assumptions can lead to seemingly inevitable and seemingly plausible conclusions about limits to material properties that are not in fact required by physical law. Examples include restrictions on Poisson's ratio based on properties of then-known materials as well as restrictions on heat capacity, elastic modulus, and thermal expansion that depend on assumptions about the material. Such assumptions may either not be stated at all or stated in a mathematical form that appears at first sight to be universally valid. Materials developed by relaxing such assumptions can have properties that are unusual or extreme; recent ones have been called metamaterials.

1.1.4 Materials vs. structures

Is it a material or a structure? An array of beams and columns in a building or a construction crane is clearly a structure. If the same geometry is scaled down to a size too small to discern with the unaided eye, it will be regarded as a material. If the size scale is a few millimeters, people may not always agree on how to classify it. There is an element of tradition in the classification. A solid block of metal is generally regarded as a material. Most metals we encounter are polycrystalline; the crystals may be on the scale of μm to millimeters or larger. The crystal structure is illustrated in old brass door handles in which the crystalline heterogeneity on a scale of millimeters to a centimeter or so is revealed to the eye by years of contact with skin which etches the metal. Even so, the metal is regarded as a material. For the present purposes, if a continuum view is taken of physical properties so that one speaks of stress, strain, elastic modulus, dielectric constant and so on, then we will regard it as a material.

1.2 Outline

This book is organized as follows.

Chapter 2 introduces basic composite structures and their elastic properties, also upper and lower bounds on properties.

Chapter 3 presents governing equations for anisotropic elasticity, and the role of material symmetry in relation to properties.

Chapter 4 introduces coupled fields in which materials can respond to multiple field variables such as those associated with elastic, electric and thermal response.

Chapter 5 presents specific materials with particulate, fibrous, and platelet inclusions and their properties.

Chapter 6 introduces cellular solids: honeycomb, foam, and lattices including structural hierarchy as well as materials with cellular structure that attain negative or extreme physical properties.

Chapter 7 presents materials of biological origin and the role of structural hierarchy in biological materials.

Chapter 8 discusses the role of the size scale of structural heterogeneity in analysis and design.

Chapter 9 presents the viscoelastic response of heterogeneous solids: creep, relaxation, energy dissipation, wave attenuation and rate dependence.

As for extreme and negative properties, the attainment of bounds on elastic moduli is presented in §2.3. Exceeding the bounds including the attainment of negative properties is presented in §2.5.1 for structural stiffness, for elastic and viscoelastic moduli in §2.5.4, for heat capacity in §2.5.6, and for electrical capacitance in §2.5.7. High values of compressive strength to density ratio in cellular materials are attained via structural hierarchy in §6.3.5. The control of Poisson's ratio including negative Poisson's ratio (auxetic materials) is presented in §6.6. Bounds on thermal expansion (§4.3.4) of two-phase composites can be exceeded by allowing void space or slip interfaces as presented in §6.7.1. The control of the sign and magnitude of the Hall effect is presented in §6.7.3. Smart materials are based on coupled fields; the principles are presented in Chapter 4.

1.3 Role of density

The rationale for considering physical properties in relation to density is as follows. The context presumes one desires a structure made of the material to be stiff and strong.

Example: effect of density in compression

Consider a column of circular cross-section, radius r and length L. Suppose the mass m is fixed but that one can choose materials of different density ρ. What is the relationship between Young's modulus E and density ρ that a designer must maximize in order to maximize the structural rigidity of the column in compression under its own weight?

Solution

The deflection u of the column is obtained from the one dimensional form of Hooke's law, $\sigma = E\epsilon$ with E as Young's modulus, σ as stress and ϵ as strain. The mass m supported a distance z down from the top of the column is $m = \rho\pi r^2 z$. The stress is the force divided by area: $\sigma = \frac{mg}{\pi r^2}$ in which g is the acceleration due to gravity. Substituting the mass, $\sigma = \rho g z$, The deflection u of the top of the column is $u = \int_0^L \epsilon dz$. The strain is $\epsilon = \frac{\sigma}{E} = \frac{\rho g z}{E}$, so $u = \frac{\rho g}{E}\frac{L^2}{2}$. The structural rigidity per mass is $\Gamma = \frac{mg/u}{m}$.

Substituting,

$$\Gamma = \frac{E}{\rho}\frac{2}{L^2}. \tag{1.1}$$

The figure of merit Γ, which is the ratio of structural rigidity to mass, is proportional to the ratio $\frac{E}{\rho}$ of elastic modulus to density for the column material. This is perhaps not surprising. However if the material is to function in bending, the figure of merit is different, as follows.

Example: effect of density in bending

Consider a rod of circular cross-section, radius r and length L, in bending. Suppose the mass m is fixed but that one can choose materials of different density ρ. What is the relationship between Young's modulus E and density ρ that a designer must maximize in order to maximize the structural rigidity of the beam in bending under its own weight?

Solution

The defection u of the beam is $u = \frac{wL^4}{8EI}$ with $I = \frac{\pi}{4}r^4$ and w as the load per length. The structural rigidity per mass is $\Gamma = \frac{rigidity}{mass} = \frac{\frac{w}{u}}{m}$ with $m = \rho\pi r^2 L$.

So $\Gamma = \frac{rigidity}{mass} = \frac{8EI}{mL^4} = \frac{8E\frac{\pi}{4}r^4}{\rho\pi r^2 LL^4} = \frac{2Er^2}{\rho L^5}$.

We are free to vary r so eliminate it from the above equation by the equation for mass and density of a cylinder.

$$\Gamma = \frac{rigidity}{mass} = \left[\frac{2m}{\pi L^6}\right]\left[\frac{E}{\rho^2}\right]. \tag{1.2}$$

The bracket on the right represents the figure of merit for the material. For tension and compression, as one might expect, the figure of merit is the ratio of stiffness to density $\frac{E}{\rho}$.

For the bending of a round rod, the figure of merit is the ratio of the modulus to the *square* of the density, $\frac{E}{\rho^2}$. We remark that for bending of a plate, the figure of merit is $\frac{E}{\rho^3}$.

1.3.1 Modulus and density

The elastic modulus of several materials is plotted vs. density in Figure 1.1. This sort of plot is called an Ashby map following similar presentations by Ashby [1]. Observe that commonly used metals have nearly the same ratio $\frac{E}{\rho}$ of modulus to density. The lower density metals have higher $\frac{E}{\rho^2}$ and $\frac{E}{\rho^3}$, the figure of merit for the bending of rods and plates respectively. Wood along the grain has favorable properties, particularly in bending, but is subject to decay as well as loss of properties in wet environments.

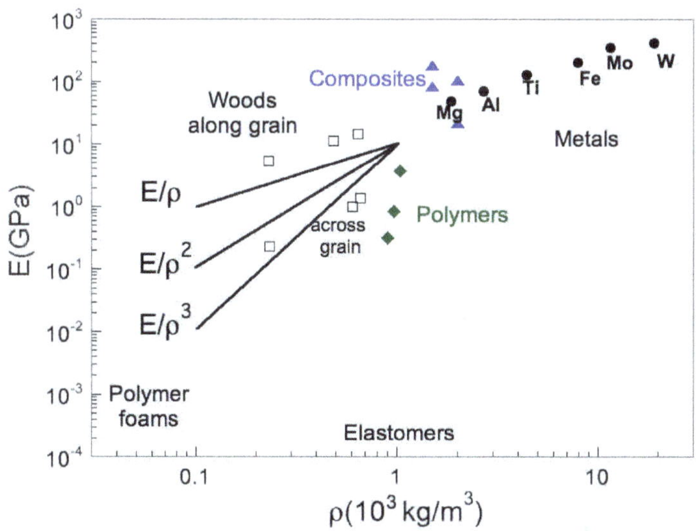

Figure 1.1: Elastic modulus vs. density for several materials, adapted from [1]. The figures of merit for structural rigidity of columns in compression and of rods and plates in compression are shown by solid lines.

1.3.2 Strength and density

The strength of several materials is plotted vs. density in Figure 1.2. Metal alloys have a considerable range of strength; the alloy constituents govern

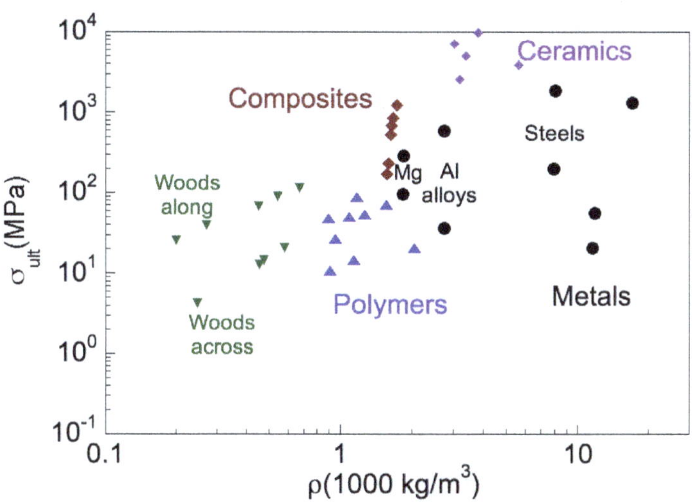

Figure 1.2: Strength vs. density for several materials adapted from [1].

the mobility of defects but do not influence the modulus much. The figure of merit for tension or compression strength is $\frac{\sigma^{ult}}{\rho}$.

For the bending [1] of rods the figure of merit is $\frac{\sigma^{ult}}{\rho^{3/2}}$ and for the bending of plates it is $\frac{\sigma^{ult}}{\rho^2}$.

1.3.3 Soft materials

For some applications the material should be compliant, not stiff. Seat cushions, protective gloves, and padding are examples. In some applications the strength should not be maximized. Crash barriers and helmets intended to absorb impact energy are in this category. The strength is tuned so that the material crushes and absorbs the energy of an impact.

Bibliography

[1] M. F. Ashby, On the engineering properties of materials, Acta Metall., 37, 1273-1293, (1989).

[2] G. W. Milton, *The Theory of Composites*, Cambridge University Press, Cambridge, U.K. (2002).

[3] J. C. Maxwell, *A Treatise on Electricity and Magnetism*, Article 314, p. 365-367, Oxford, Clarendon Press, Oxford, U.K. (1873).

Chapter 2

Structures, properties, bounds

2.1 Introduction

The properties of composites are greatly dependent upon *microstructure*. Composites differ from nominally homogeneous materials in that considerable control can be exerted over the larger scale structure, and hence over the desired properties. In particular, the properties of a composite depend upon the *shape* of the heterogeneities, upon the *volume fraction* occupied by them, and upon the *interface* between the constituents. Volume fraction refers to the ratio of the volume of a constituent to the total volume of a composite specimen. *Cellular solids* (Chapter 6) are those in which the 'inclusions' are voids or cells, filled with air or liquid. Cellular solids include honeycombs, in which the structure is largely two-dimensional, and foams, in which the structure is fully three-dimensional. Foams can be open-cell, in which the foam has a structure of 'ribs', with no barrier between adjacent cells; or closed-cell, in which plate or membrane elements separate adjacent cells. Open-cell foams are of particular interest in the context of viscoelasticity [1], since viscoelastic damping can arise as a result of the viscosity of fluids (such as water or air) moving through the pore structure. *Hierarchical* composites have structural elements within structural elements. The *coated spheres* morphology has multiple length scales. The entire volume of the material is filled with particles of one phase coated with a layer of a second phase. This morphology is used in theoretical analyses of extremal behavior: the attainment of bounds, §2.3. A detailed mathematical treatment of structure property relations, the analysis of bounds and the attainment of bounds has been thoroughly presented [2]. More extreme behavior, the exceeding of bounds, is possible if one relaxes assumptions made in the derivation of bounds: §2.5.

The shape of heterogeneities in a composite is classified as follows. The principal inclusion shape categories (Figure 2.1) are the particle, with no long dimension; the fiber, with one long dimension; and the platelet (flake, lamina), with two long dimensions. The inclusions may vary in size and shape within a category. For example, particulate inclusions may be spherical, ellipsoidal, polyhedral, or irregular.

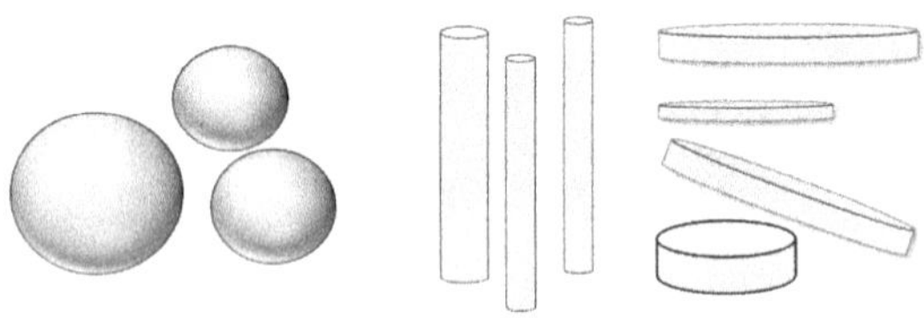

Figure 2.1: Idealized composite inclusion shapes: left, particles; center, fibers; right, platelets.

2.2 Bounds on properties

2.2.1 Bounds on elastic constants of a homogeneous solid

Bounds on elastic moduli

For a homogeneous, isotropic elastic material, the bulk modulus K (also called B) and the shear modulus G are shown to be positive based on positive definite strain energy [3] [4]. The positive definite energy concept is used to demonstrate that other physical properties are positive [5]. For example the compressibility (inverse bulk modulus) is shown to be positive as follows [5]. The first law of thermodynamics is

$$dQ = dU + dW, \tag{2.1}$$

with dQ as an increment of heat going into the material, dU as an increment of energy and dW as the work done by the material. The second law may be written

$$dQ = TdS, \tag{2.2}$$

with T as temperature and S as entropy. The pressure P at constant entropy S as indicated by the subscript is, from $dW = PdV$,

$$P = -\frac{\partial U}{\partial V}|_S \tag{2.3}$$

with V as volume. If the material is to be stable, the internal energy must be a minimum with respect to all variations $d^2U > 0$.

$$d^2U = \frac{\partial^2 U}{\partial^2 S}|_V (dS)^2 + 2\frac{\partial^2 U}{\partial S \partial V}|_{VS}(dS)^2 + \frac{\partial^2 U}{\partial^2 V}|_S (dV)^2. \qquad (2.4)$$

Constant volume V and constant entropy S are indicated by subscripts. So from Equation 2.3, considering only variations in volume at constant entropy,

$$-\frac{\partial P}{\partial V}|_S = \frac{1}{Vk_S} > 0 \qquad (2.5)$$

with $k_S = \frac{1}{V}\frac{\partial V}{\partial P}_S$ as the adiabatic compressibility. So the compressibility, hence the bulk modulus K, is positive. By similar arguments via positive definite strain energy, the shear modulus G is also positive. So the bounds on elastic moduli based on positive definite strain energy are

$$K > 0, G > 0. \qquad (2.6)$$

Bounds on Poisson's ratio

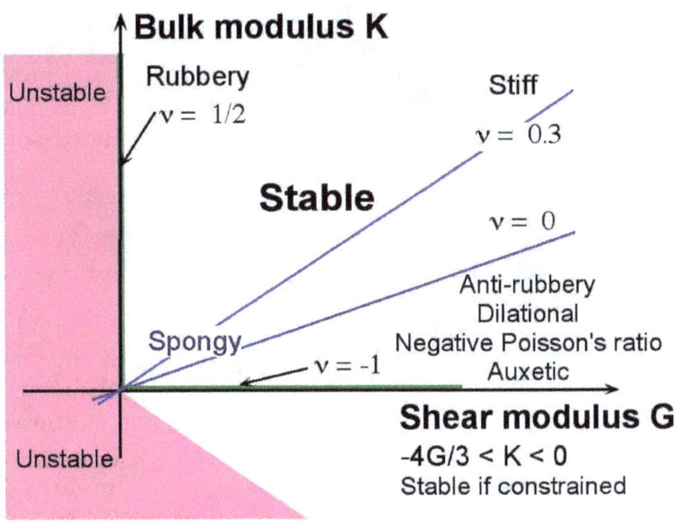

Figure 2.2: Map of bulk modulus K vs. shear modulus G, showing regions of stability and Poisson's ratio ν, adapted from [8].

In isotropic elastic materials, the elastic constants are interrelated. Positive bulk modulus and positive shear modulus imply bounds on Poisson's ratio. The allowable range of Poisson's ratio for 3D isotropic stable materials is therefore

$$-1 < \nu < 0.5. \tag{2.7}$$

Poisson's ratio is a cross property with no associated energy, so negative values are admissible as is now well known in designed cellular solids [6] (§6.6), in some materials during phase transformation, and in designed lattices. The allowable range of Poisson's ratio was at one time a matter of debate as discussed in §3.7.2. A seemingly plausible model of interaction between atoms predicted a Poisson's ratio of 1/4 for all materials. As we know, that is overly restrictive.

The stability region for positive moduli may be mapped [7] to include negative Poisson's ratio as shown in the upper right quadrant of Figure 2.2. The stability region in the map is extended for constrained materials with a negative bulk modulus [8].

The stability criterion of positive definite strain energy implies stability of an object with free surfaces with zero applied stress. The stability criterion for an object constrained at its surfaces is that of strong ellipticity [3], a less stringent condition. The shear modulus G and the constrained modulus C (see §3.3.5) must be positive for strong ellipticity. Strong ellipticity entails real shear wave velocity and real longitudinal wave velocity.

But $C = K + \frac{4}{3}G$ following Equation 3.18. So a material with a negative bulk modulus K can be stable provided it is constrained and provided the bulk modulus is not too negative as shown in the lower right quadrant of Figure 2.2. The range of isotropic Poisson's ratio associated with strong ellipticity is [9]

$$-\infty < \nu < 0.5. \tag{2.8}$$

or

$$1 < \nu < \infty. \tag{2.9}$$

A material that does not obey strong ellipticity becomes unstable and exhibits bands of heterogeneity [10]. Indeed such bands are observed in materials that have undergone particular phase transformations, e.g. barium titanate below the cubic to tetragonal transformation temperature as shown in Figure 4.3.

Composites with partially constrained inclusions of negative modulus allow attainment of extreme properties as discussed in §2.5.4.

2.2.2 Bounds on heat capacity of a homogeneous solid

The heat capacity is the ratio of heat change to temperature change,

$$C_V = \frac{\partial Q}{\partial T}|_V, \tag{2.10}$$

in which the subscript indicates constant volume.

This may be written in terms of the entropy or the energy

$$C_V = T\frac{\partial S}{\partial T}|_V = \frac{\partial U}{\partial T}|_V. \tag{2.11}$$

Again [5] invoking $d^2U > 0$,

$$\frac{\partial T}{\partial S}|_V = \frac{T}{C_V} > 0. \tag{2.12}$$

The temperature $T > 0$ so $C_V > 0$. The heat capacity is positive which means if heat is added, the temperature increases.

2.2.3 Bounds on composite elastic properties

Bounds on composite properties [2] can be useful because composite microstructure can be sufficiently complicated that exact analysis of the properties is difficult. The simplest bounds are the Voigt and Reuss formulae. These equations were originally introduced for calculating the elastic moduli of an aggregate of crystals. Such a calculation is pertinent because almost all metals and ceramics and some polymers in common use are polycrystalline. The Voigt formula assumes the strain is uniform in the polycrystalline material; the Reuss formula assumes the stress is uniform. It was found that the measured moduli for polycrystalline materials always lie between the Reuss and Voigt values. Hill showed via an energy method that the Voigt and Reuss formulae represent bounds on the modulus of a polycrystalline material [11] and of a two-phase composite [12]. The bounds for an isotropic composite apply to the shear modulus and to the bulk modulus.

Voigt bound

For elastic materials with no slip between the phases, and assuming two phases, the Voigt bound is

$$G_c = G_1V_1 + G_2V_2 \tag{2.13}$$

in which G_c, G_1 and G_2 refer to the shear modulus of the composite, phase 1 and phase 2, and V_1 and V_2 refer to the volume fraction of phase 1 and phase 2 with $V_1 + V_2 = 1$. The dependence of stiffness on volume fraction is shown in Figure 2.3.

Reuss bound

For a material containing two elastic constituents, the Reuss bound is

$$\frac{1}{G_c} = \frac{V_1}{G_1} + \frac{V_2}{G_2}. \tag{2.14}$$

The dependence of stiffness on volume fraction is shown in Figure 2.3. In the compliance formulation, the Reuss relation can be written more simply in terms of the compliances $J = 1/G$.

$$J_c = J_1 V_1 + J_2 V_2. \tag{2.15}$$

Hashin-Shtrikman bounds

The upper and lower bounds of *stiffness* of two-phase and multi-phase composites have been obtained by Hashin, and by Hashin and Shtrikman in terms of volume fraction of constituents [13], [14]. Much work on composites is reviewed by Hashin [15]. Allowing for 'arbitrary' phase geometry, the upper and lower bounds on the elastic moduli of an *isotropic* composite as a function of composition were developed using variational principles. The lower bound for the elastic shear modulus G_L of the composite was given, via variational energy arguments, as [14]

$$G_L = G_2 + \frac{V_1}{\frac{1}{G_1 - G_2} + \frac{6(K_2+2G_2)V_2}{5(3K_2+4G_2)G_2}}, \tag{2.16}$$

in which K_1, G_1, V_1; K_2, G_2 and V_2 are the bulk modulus, shear modulus and volume fraction of phases 1 and 2, respectively. Here $G_1 > G_2$, so that G_L represents the lower bound on the shear modulus. Interchanging the numbers 1 and 2 in Equation 2.16 results in the upper bound G_U for the shear modulus. Bounds on the stiffness are shown in Figure 2.3. The lower bound for the bulk modulus of an elastic material is as follows [15].

$$K_L = K_2 + \frac{V_1(K_1 - K_2)(3K_2 + 4G_2)}{(3K_2 + 4G_2) + 3V_2(K_1 - K_2)} \tag{2.17}$$

Here K_1 and K_2 are the bulk moduli of the two phases, G_1 and G_2 are the shear moduli, and V_1 is the volume fraction of the first phase. As for Poisson's ratio, it is related to the moduli by $\nu = \frac{3K-2G}{6K+2G}$, assuming isotropy. In the viscoelastic case, these moduli become complex: $K_1 \rightarrow K_1^*$.

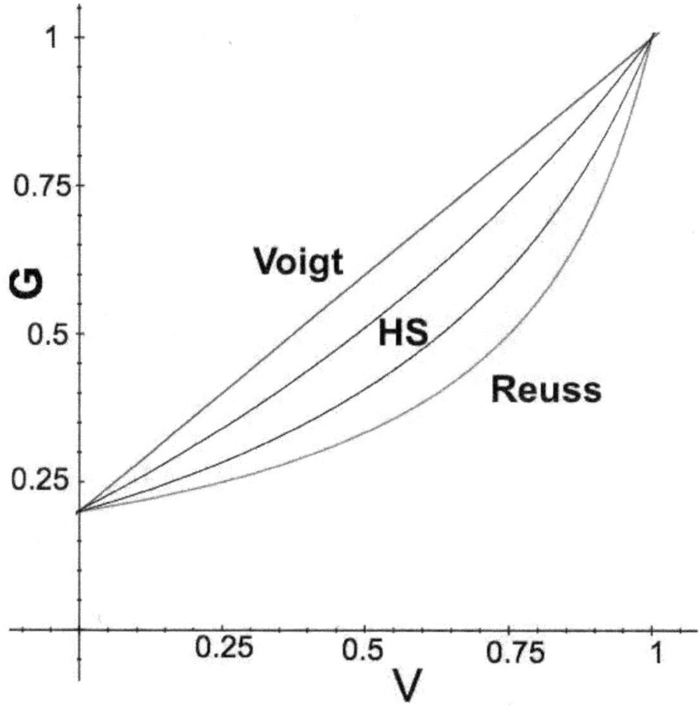

Figure 2.3: Bounds on shear modulus G vs. volume fraction V of a constituent: Voigt, Reuss and Hashin-Shtrikman. One constituent is assumed to be five times stiffer than the other constituent. HS refers to the Hashin-Shtrikman bounds for isotropic composites; Poisson's ratio is assumed to be 0.33.

2.2.4 Bounds on composite dielectric constant

The dielectric constant k of a two-phase composite is subject to bounds based on energy arguments. The Voigt and Reuss relations constitute bounds on the dielectric or conductivity behavior of a two-phase composite of any structure. The Voigt bound is

$$k_c = k_1 V_1 + k_2 V_2 \tag{2.18}$$

and the Reuss bound is

$$\frac{1}{k_c} = \frac{V_1}{k_1} + \frac{V_2}{k_2}. \tag{2.19}$$

The subscripts c for the composite and 1 and 2 for the constituents and for the volume fractions are defined as in the elastic case.

2.3 Attaining the bounds on properties

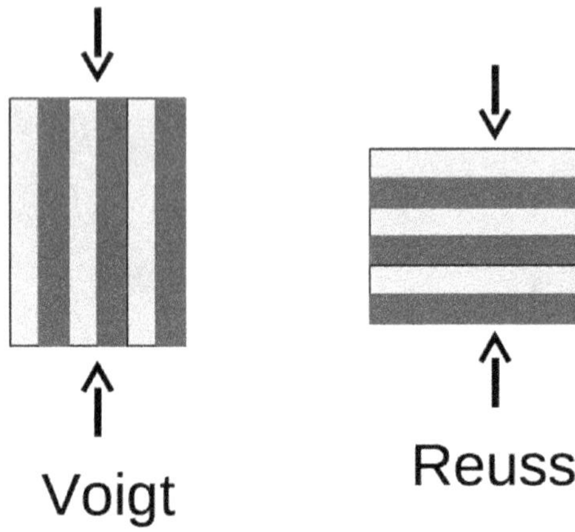

Figure 2.4: Voigt and Reuss structures consist of laminations of two solid phases. Forces are indicated by arrows.

Composite structures such as those of the Voigt and Reuss composites give rise to exact analytical solutions for the elastic moduli. Similarly, properties of composites containing dilute concentrations of spherical or platelet inclusions are also known in terms of constituent properties. Moreover, for the simplest case of an *elastic* two-phase composite, the stiffness of Voigt and Reuss composites (Figure 2.4) described below represent rigorous upper and lower bounds on the modulus for a given volume fraction of one phase (Figure 2.3). The Hashin-Shtrikman equations presented below represent upper and lower bounds on the elastic stiffness of *isotropic* composites with two phases or constituents. The assumptions underlying the bounds include perfect bonds between the constituents, and that the energy density is positive definite. That means that each constituent is in its lowest energy state.

Bounds are useful in that one can predict the range of possible behavior in materials for which the microstructure is too complex for an analytical determination of properties. If the constituent moduli have a large ratio then the bounds are far apart, reducing their utility.

2.3.1 Voigt composite

The Voigt composite can contain laminations as shown in Figure 2.4; the strain in each phase is the same. The Voigt formula applies to the shear modulus G. The Voigt formula also applies for Young's modulus E provided each phase has the same Poisson's ratio. If one applies axial load in the fiber direction of a unidirectional fibrous composite, the Young's modulus also follows the Voigt formula provided the Poisson's ratio of each phase is the same. The Voigt formula is also called the rule of mixtures, but it is not a general rule, and it is only applicable to a particular structure and stress direction. The analysis can be generalized to more than two phases provided they are aligned in the same orientation.

Example

Derive $E_c = E_1V_1 + E_2V_2$, for the elastic Voigt composite microstructure, in one dimension, neglecting Poisson effects. What is the situation if Poisson effects are considered? Is the Voigt formula exact in that case? If so, for what values of Poisson's ratio? What will the composite modulus look like if the constituents are linearly viscoelastic, describable by complex moduli?

Solution

Since it is elastic, Hooke's law for phases 1 and 2 is

$$\sigma_1 = E_1\epsilon_1, \text{ and } \sigma_2 = E_2\epsilon_2.$$

The loads F in terms of the cross-sectional areas A are

$$F_1 = A_1\sigma_1 = E_1\epsilon_1A_1, \text{ and } F_2 = A_2\sigma_2 = E_2\epsilon_2A_2.$$

In view of the geometry, the load F_c on the composite block must be the sum of the loads on the constituents.

$$F_c = F_1 + F_2,$$

$F_c = \sigma_cA_c = \sigma_1A_1 + \sigma_2A_2$, so, dividing by A_c, and recognizing the volume fraction for this geometry as the ratio of cross-sectional areas,

$$\sigma_c = \sigma_1\tfrac{A_1}{A_c} + \sigma_2\tfrac{A_2}{A_c} = \sigma_1V_1 + \sigma_2V_2.$$

Finally, divide by the strain, which is the same in both constituents provided they are perfectly bonded.

$$E_c = E_1V_1 + E_2V_2,$$

as desired.

As for Poisson effects, the transverse strain will be the same in each phase if the constituent Poisson's ratios are equal. For that case there are no transverse stresses, so the Voigt formula for E is exact for equal Poisson's ratios.

2.3.2 Reuss composite

The geometry of the Reuss model structure is shown in Figure 2.4; each phase experiences the same stress but different strain. Since the constituents are aligned, this composite is *structurally* anisotropic. Since the Reuss laminate is identical to the Voigt laminate except for orientation with respect to the stress, the laminate is *mechanically* anisotropic. If the

Reuss model is considered in three dimensions (Figure 2.5), it applies to Young's modulus E provided Poisson's ratio is zero for both phases. The Reuss model applies to the shear modulus G; there are no Poisson effects in shear.

2.3.3 Laminates: dielectric constant

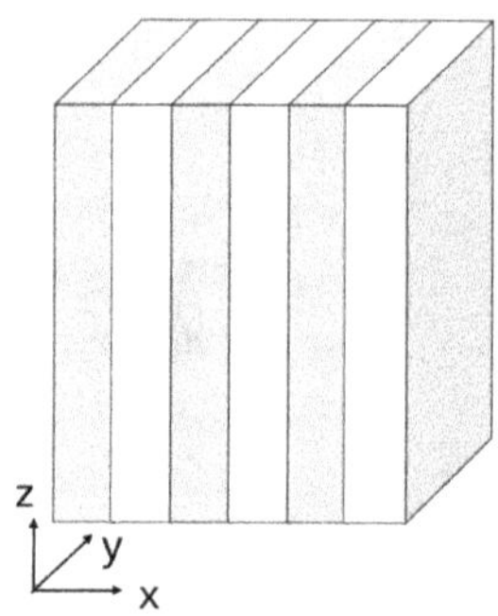

Figure 2.5: Rank 1 three-dimensional laminate.

To determine the dielectric constant of a three-dimensional laminate, recall the relation for adding capacitances C_1 and C_2 in series: $C_{ser}^{-1} = C_1^{-1} + C_2^{-1}$. The corresponding relation for adding capacitances in parallel is $C_{par} = C_1 + C_2$. These are shown using the definition of capacitance: charge q is related to voltage V via the capacitance C: $V = q/C$. The parallel relation arises from the addition of charge on adjacent capacitors. In series the charge on each capacitor is the same, and the total voltage is the sum of voltages on each capacitor.

The capacitance of a capacitor is related to the relative dielectric permittivity or dielectric constant k of the material comprising the capacitor, $C = \epsilon_0 k \frac{A}{h}$ with $\epsilon_0 = 8.85 \times 10^{-12}$ farad / m as the permittivity of space, A is the area, and h as the thickness.

For an electric field in the z direction upon the laminate shown in Figure 2.5, the voltage across each lamina is the same and the charges are additive. So $C_c = C_1 + C_2$. The thickness h_z is that of the laminate in the z direction and is the same for each lamina. The areas are additive, so $A_c = A_1 + A_2$.

So $\frac{k_c A_c}{h_z} = \frac{k_1 A_1}{h_z} + \frac{k_2 A_2}{h_z}$.

But the volume fraction of phase 1 is $V_1 = \frac{A_1}{A_c}$ and $V_2 = \frac{A_2}{A_c}$.

So a Voigt relation applies for an electric field in the z direction, $k_c = k_1 V_1 + k_2 V_2$.

For a field applied in the y direction, the geometry is identical so a Voigt relation also applies.

For a field applied in the x direction, the capacitances are in series so $\frac{1}{C_c} = \frac{1}{C_1} + \frac{1}{C_2}$.

The area of each lamina is the same but the thickness values differ. Assume there are n laminae of each type; n is factored out. $\frac{h_c}{k_c A} = \frac{h_1}{k_1 A} + \frac{h_2}{k_2 A}$.

The volume fraction of phase 1 is $V_1 = \frac{h_1}{h_1+h_2}$; similarly for phase 2, $V_2 = \frac{h_2}{h_1+h_2}$. Divide the above by $h_c = h_1 + h_2$.

So, in the x direction, the permittivity follows a Reuss relation: $\frac{1}{k_c} = \frac{V_1}{k_1} + \frac{V_1}{k_2}$.

By determining the laminate permittivity in the three principal directions, the dielectric behavior of the laminate is fully characterized. The solution for dielectric permittivity is exact. Moreover, the Voigt and Reuss relations represent rigorous bounds on the dielectric or conductivity behavior of a two-phase composite of any structure [16]. If the composite is isotropic, more stringent Hashin-Shtikman type bounds apply. The bounds are attainable via laminates.

The permittivity is governed by a tensor of second rank (§3.6); the conductivity also is a second rank property with a governing equation similar to that of the dielectric permittivity. The analysis therefore applies to conductivity as well. This is in contrast with elasticity which uses a tensor of fourth rank. The dielectric case has no counterpart of Poisson's ratio that occurs in the elastic case.

2.3.4 Laminates: structural hierarchy

The Voigt and Reuss composites are elementary laminates that can be generalized by considering such composites in three dimensions as shown in Figure 2.5 or via structural hierarchy. Laminates can be made hierarchical by making laminae themselves laminated as shown in Figure 2.6. The rationale is to allow more freedom in the structure to allow control of the physical properties. At each level of the hierarchy one may assume a different volume fraction, hence ratio of layer thickness. This freedom is helpful in developing bounds on behavior or in conceptualizing materials with extremal properties. Laminates that themselves contain laminates were envisaged in 1873 by James Clerk Maxwell [21] in the context of conductivity analysis. These laminates have hierarchical structure. The laminate of Maxwell was third rank as shown in Figure 2.7. The rank of a laminate is the number of widely spaced length scales used to construct it [2]. Bounds have been derived [22] more recently for the conductivity of polycrystalline materials using similar hierarchical laminates. Hierarchical laminates also enable attainment of bounds on dielectric permittivity or conductivity [16].

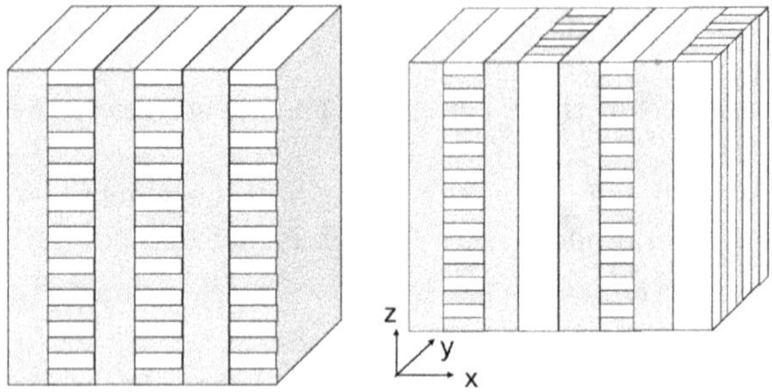

Figure 2.6: Left, rank 2 hierarchical laminate. Right, rank 2 hierarchical laminate with alternating rotated laminae.

Explicit forms for the properties of laminates of arbitrarily high rank were given [16].

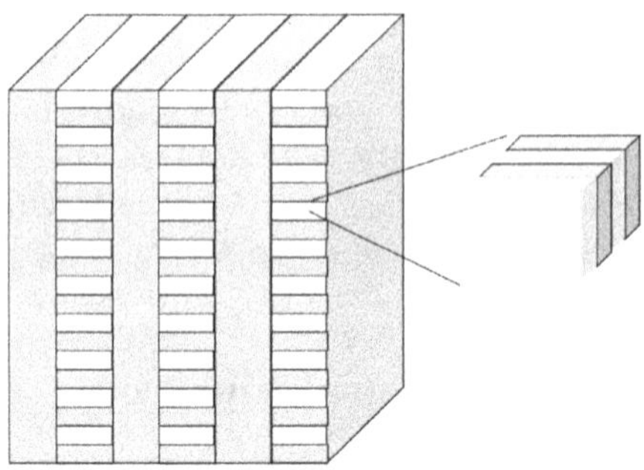

Figure 2.7: Rank 3 hierarchical laminate.

As for elasticity, the rank 1 laminate in Figure 2.5 is Voigt or Reuss depending on the direction of the stress. Analysis of the laminates in Figure 2.6 and Figure 2.7 is simplified by the fact that the laminae are orthogonal. One may begin with the smallest laminae, apply the Voigt or Reuss equation depending on the direction of the load, then assume the next larger size laminae consist of a continuum with properties just calculated. The process is then repeated until the full hierarchical laminate has been analyzed. The procedure using elementary formulae is warranted only if the Poisson's ratio of each phase is zero; rigorous analysis has been done for general Poisson's ratio. A similar procedure may be done rigorously with the dielectric constant of hierarchical laminates via the notion of adding capacitances in series or parallel. Similarly, conductivity is represented by a second rank tensor and is subject to analysis via laminates. The rank 2 laminate with alternating laminae, Figure 2.6 right, is helpful [2] in the analysis of conductivity in that it allows the use of a lower rank in obtaining bounds.

Hierarchical laminates to attain bounds on elastic moduli generally contain laminae in oblique directions as discussed below.

2.3.5 Attaining the Hashin-Shtrikman bounds: spheres

The Hashin-Shtrikman formulae represent exact solutions for moduli of isotropic hierarchical composites. As for attainment of bounds on elastic properties, the bounds on bulk modulus can be attained by a hierarchical coated sphere composite morphology, Figure 2.8. The coated sphere is a neutral inclusion: it does not disturb the assumed hydrostatic stress state in the surrounding medium, which is assumed to have a bulk modulus equal to that of the coated inclusion. Since the stress and strain fields are undisturbed in the process, the loads and displacements at the boundary are undisturbed, hence the effective bulk modulus of the medium remains unchanged by the progressive addition of inclusions [19]. For the lower bound, phase 1 is in the core, and phase 2 is in the spherical shell. The demonstration of this is done via the theory of elasticity. The coated

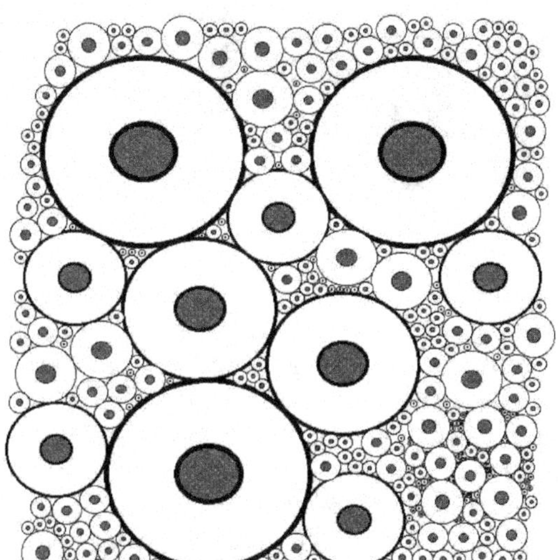

Figure 2.8: Coated sphere composite structure.

sphere structure has been suggested [2] to have a shear modulus that depends on the detailed arrangement of the spheres.

There are also bounds for properties of three phase composites. Analytical approaches to attaining them are sufficiently complicated that numerical topology optimization procedures have been used to solve the inverse homogenization problem [20]; three phase composites have been found with a maximal effective bulk modulus.

2.3.6 Attaining the Hashin-Shtrikman bounds: laminates

Exact attainment of the Hashin-Shtrikman shear modulus formula is possible via a hierarchical laminate morphology [17] [23], Figure 2.9. The Hashin-Shtrikman bounds for bulk and shear modulus are simultaneously achieved through a finite number of layering processes in hierarchical laminates [18]. The laminae have oblique directions. In summary, the Hashin-Shtrikman bounds for isotropic two-phase composites are based on an energy principle. The bounds are attainable via hierarchical structures. The Hashin- Shtrikman lower formula represents an approximation for composites with a dilute concentration of stiff spherical inclusions.

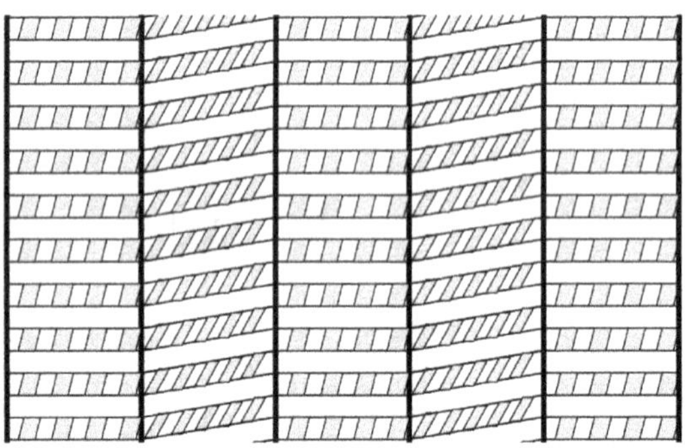

Figure 2.9: Hierarchical laminate used to attain shear modulus bound, adapted from [17].

Physical properties such as the elastic modulus may be represented as tensors, as presented in Chapter 3. For isotropic materials there are two independent elastic constants; for anisotropic materials, there can be as many as 21 independent elastic constants. A variety of elastic tensors are realizable via laminates [24]. One may approach bounds on properties if there is sufficient contrast between the moduli of the constituents. The limit of a constituent stiffness tending to zero corresponds to void space.

Indeed, materials with void space have been developed to exhibit extreme properties as discussed in §6.6 and §6.7.1.

2.4 Inclusion shape: dilute concentration

2.4.1 Spherical inclusions

For a small volume fraction $V_2 = 1 - V_1$ of spherical elastic inclusions (particles) in a continuous phase of another elastic material, matrix phase 1, the shear modulus of the isotropic composite G_c was given as [25]

$$\frac{G_c}{G_1} \approx 1 - \frac{15(1-\nu_1)(1-\frac{G_2}{G_1})V_2}{7-5\nu_1+2(4-5\nu_1)\frac{G_2}{G_1}} \tag{2.20}$$

in which ν_1 is the Poisson's ratio of phase 1, and phase 1 and phase 2 represent the matrix material and the inclusion material respectively. The demonstration of this is done via the theory of elasticity. The stiffness of such a composite, as a function of the volume fraction of inclusions, is close to the Hashin-Shtrikman lower bound for isotropic materials.

For larger volume fractions of inclusions, analysis is more complicated as a result of the interaction of stress fields around nearby inclusions. Stiff spherical inclusions are less efficient, per volume, in achieving a stiff composite than fiber or platelet inclusions. Conversely, soft spherical inclusions have the least effect in reducing the stiffness, in comparison to other inclusion shapes. The case of a large concentration of spherical inclusions is not amenable to simple analytical solutions, so finite element analyses are done [26].

2.4.2 Fiber inclusions

For a three-dimensional dilute concentration of *randomly oriented* fiber elastic inclusions of phase 2 in a matrix of phase 1, the Young's modulus E_c of the composite is as follows [27], assuming a Poisson's ratio of both fiber and matrix to be 1/4. Random orientation gives rise to a composite which is isotropic.

$$E_c \approx \frac{1}{6}E_2V_2 + E_1\frac{1+\frac{1}{4}V_2+\frac{1}{6}V_2^2}{1-V_2}. \tag{2.21}$$

By contrast if all the fibers are aligned in the same direction, the modulus for deformation along the fiber direction follows the Voigt relation, Equation 2.13, so the aligned composite is considerably stiffer than one with random fibers. As for dilute random fibers in two dimensions, $E_c \approx \frac{1}{3}E_2V_2$, so this is stiffer than the case of three-dimensional random orientation.

2.4.3 Platelet inclusions

For a dilute concentration of *randomly oriented* platelet elastic inclusions (flakes) of phase 2 in a matrix of phase 1, the shear modulus G_c of the composite was given as follows [25].

$$E_c \approx E_1 V_1 + \frac{1}{2} E_2 V_2. \tag{2.22}$$

The random orientation of constituents gives rise to a composite which is isotropic. Stiff platelets are more efficient, per volume, in achieving a stiff composite than fiber or particle inclusions. At low volume fraction V_2 of the randomly oriented platelets, the stiffening effect of the platelets is $\frac{1}{2}$ of the value obtained in the Voigt composite (neglecting Poisson effects).

The predicted stiffness of a random platelet reinforced composite is identical to the Hashin-Shtrikman upper bound for isotropic materials [28], however this limiting stiffness will only be achieved if the platelets are infinitely thin [29]. A composite of randomly oriented platelet inclusions attains the Hashin-Shtrikman bounds for bulk and shear modulus [30] [2].

2.5 Exceeding bounds

Bounds on properties are demonstrated using assumptions that are reasonable and may seem inevitable. Some such assumptions entail restrictions on the type of physical system considered. If one relaxes appropriate assumptions by choice of a material or system that does not obey them, one may attain negative or extreme physical properties. Attainment of extreme or negative Poisson's ratios has been done with shaped cellular solids, §6.6; similarly extreme or negative thermal expansion, §6.7.

2.5.1 Negative structural stiffness and extreme damping

The stiffness of a structural element may be represented by a spring constant k. The energy is $U = \frac{1}{2}kx^2$ in which x is the displacement. Assuming the energy is positive definite, the stiffness k must be positive. Positive definite strain energy, used in demonstrating positive properties, means that the system is initially in its minimum energy state. One can certainly prepare a system with stored energy. Strain energy is stored during buckling; by allowing a pre-strain, negative structural stiffness (spring constant) is possible [31]. The curve of force vs. displacement for some systems exhibits a reversal in slope as illustrated in Figure 2.10. Systems that buckle in this

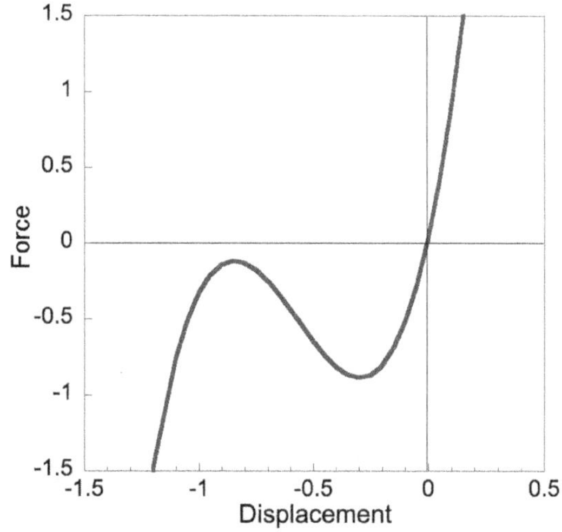

Figure 2.10: Force vs. displacement for a system that undergoes buckling in compression (negative displacement); arbitrary units of force and displacement.

way include flexible columns with flat ends, flexible tubes, and tetrakaidecahedra made with flexible ribs. The structural stiffness for small increments of displacement in the reverse slope region is negative. The system may be stabilized in that region by a constraint as in Figure 2.11. Arrows indicate applied imposed displacement. If, rather than controlling the displacement, the force is controlled, the system may not be stable.

If one combines positive and negative stiffness in a Voigt configuration, Equation 2.13, the overall stiffness can be made small. That can be useful in vibration isolation systems in which a compliant support is beneficial. An ordinary spring which is compliant will be very long. It may be more practical to neutralize some of the stiffness of an ordinary spring with a negative one. Such systems are used in practical vibration isolators [32]. These isolators use elastic structures loaded to approach elastic instability, giving rise to a negative stiffness component.

If one combines positive and negative stiffness in a Reuss configuration, Equation 2.15, positive and negative values sum to a small or zero value in the denominator. That suggests the overall stiffness can be made large. Large positive stiffness in such a system is, however, unstable. Positive and negative structural stiffness combined near the stability limit can be used to attain extremely high mechanical damping (§9.2.3) in discrete systems [33]. In an experiment, the instability of such a system is stabilized via servo control of the displacement.

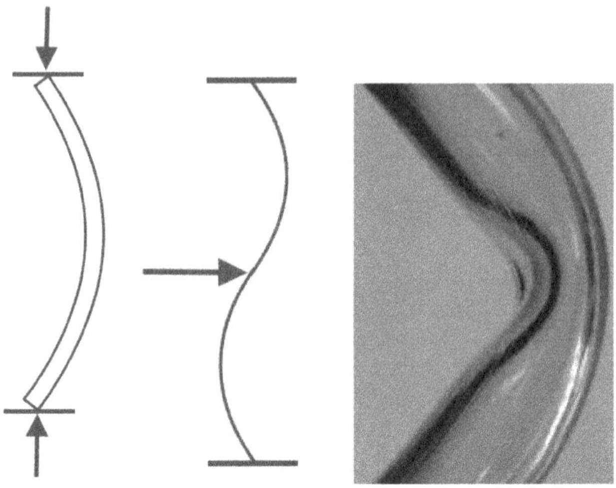

Figure 2.11: Systems that exhibit negative structural stiffness. Left: Buckling of a column with ends that tilt. Center: Constrained sigmoid column pressed laterally. Right: Tube buckling. Arrows indicate imposed displacement.

2.5.2 Extreme nonlinear energy dissipation

Negative stiffness constituents contribute instability. If the instability occurs over a sufficient range of deformation, the overall response becomes nonlinear. Instability associated with post-buckling behavior of columns (Figure 2.11, left) can give rise to extremely large energy dissipation combined with high stiffness [34] without damage. Small dampers of this type provided significant damping of the bending vibration of a metal rod [35]. A helicoidal bistable shell was developed to envelop a solid cylinder; breakages were considered in the context of energy dissipation [36].

2.5.3 Phase transformations

Phase transformations are of interest in the context of achieving extreme physical properties such as large piezoelectric sensitivity (§4.2.3) and large values of other coupled field effects such as the piezocaloric effect (§4.3.6), as well as extreme elastic and viscoelastic behavior in composites (§2.5.4). Also, piezoelectric materials (§4.2.2) may cease to be piezoelectric following a phase transformation.

As temperature or pressure changes, materials can undergo phase transformations from one internal organization to another. For example, water can exist as a liquid, or as a gas (steam) or as ice. Phase transformations can occur from one solid phase to another; the organization of the material at the interatomic scale undergoes a change analogous to buckling. Elastic moduli can soften considerably in the vicinity of a phase transformation as shown in Figure 2.12. If softening of the bulk modulus dominates as is the case for the materials shown, the Poisson's ratio is negative over a range of temperatures. The change in the crystal structure may resemble the folded-in structure observed earlier in designed negative Poisson's ratio foams as discussed in §6.6. Moduli do not soften all the way to zero because there are heterogeneities in the environmental temperature or in the material specimen. Similarly, a polymer gel exhibits volume transition

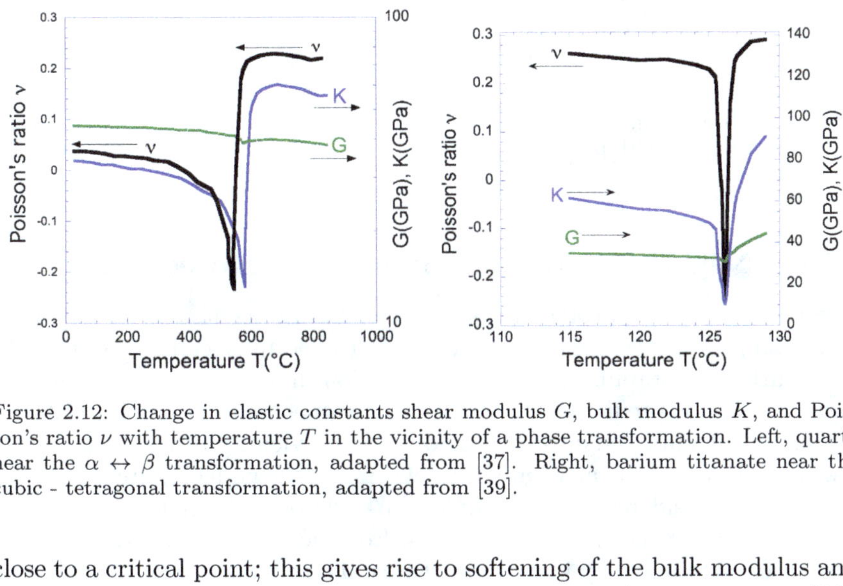

Figure 2.12: Change in elastic constants shear modulus G, bulk modulus K, and Poisson's ratio ν with temperature T in the vicinity of a phase transformation. Left, quartz near the $\alpha \leftrightarrow \beta$ transformation, adapted from [37]. Right, barium titanate near the cubic - tetragonal transformation, adapted from [39].

close to a critical point; this gives rise to softening of the bulk modulus and a Poisson's ratio that is negative over a narrow range in temperature [38].

Negative elastic moduli are anticipated in the context of Landau's [53] theory of phase transformation. As temperature T is lowered, an energy function of strain and temperature initially has a single minimum, then it gradually flattens, then it develops two minima as shown in Figure 2.13. The slope of the energy curve is proportional to the stiffness or elastic modulus. In the absence of constraint, one or more moduli may soften but will remain positive in the regime of stability. The crystal manifests instability by changing to a new crystal form. For a shear strain, the transformation is martensitic; for a hydrostatic strain, the transformation is a volume change transformation.

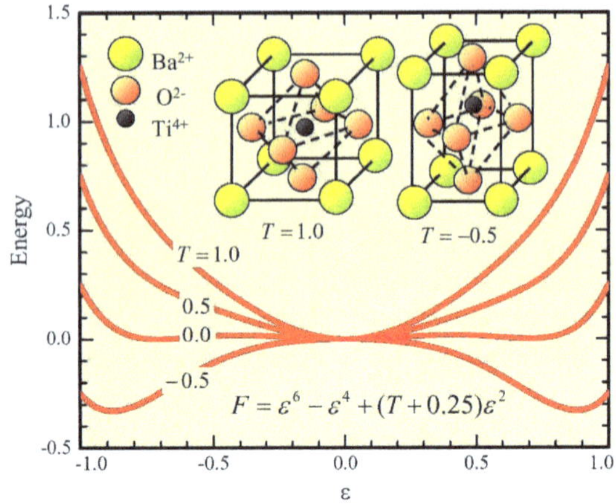

Figure 2.13: Energy density vs. strain ϵ and temperature T associated with a phase transformation [40], with permission.

2.5.4 Negative and extreme moduli: stored energy

Negative moduli are forbidden based on energy arguments related to stability (§2.2.1). A material with a negative elastic modulus can be constrained at its surface to stabilize it if it is an inclusion in a composite [41]. Such a composite [41] is predicted to exhibit extreme mechanical damping (§9.3.8) as well as extremely high or low moduli in the vicinity of the balance between positive and negative modulus effects, as shown in Figure 2.14. The moduli are considered to be complex in a viscoelastic material. Moduli corresponding to the Hashin-Shtrikman Equation 2.16 used for such composites are attainable by known structures. The curves in Figure 2.14 have a formal resemblance to resonant behavior but there are no inertia terms. The reason is the Hashin-Shtrikman equations have two terms in the denominator that have opposite signs if constituent moduli are allowed to be negative. The equation for compliance of a system with inertia also has two terms in the denominator with very different physical meaning: the inertia term has an opposite sign to the restoring force term.

Extreme modulus and damping effects have been observed [42] [40] in composites with inclusions in the vicinity of a phase transformation. In a composite, the phase transformation of an inclusion is partially constrained, so that energy is stored. Such a composite does not satisfy the assumption of positive definite strain energy assumed in bound analyses.

Composites with phase transforming inclusions in the negative stiffness regime can be stiffer than diamond [40] over a range of temperature during

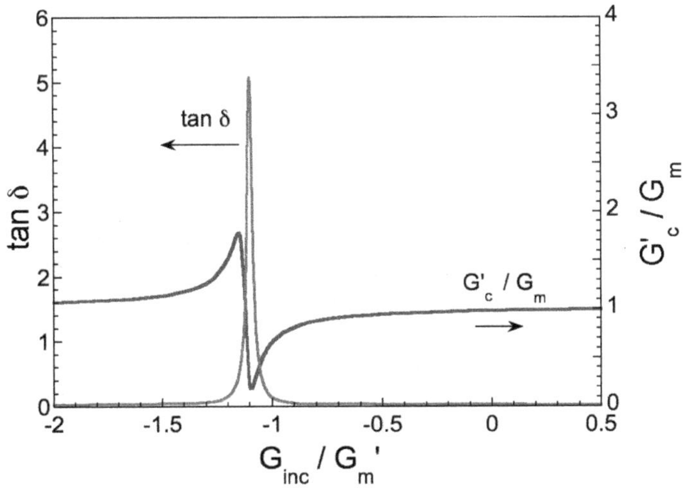

Figure 2.14: Shear modulus G'_c (real part) of composite and mechanical damping tan δ, vs. inclusion shear modulus G_{inc} normalized to matrix shear modulus G_m. The composite contains 1% by volume of spherical particulate inclusions. Assumed matrix damping is tan $\delta = 0.025$. Adapted from [42].

a temperature scan.

Composites with negative stiffness inclusions can be stable [43] [44] [45] provided the bulk modulus of the inclusions is not too negative. Further stability analysis showed elastic composites with a negative-stiffness phase can exhibit stable extremely high effective dynamic but not stable static stiffness [46]; stabilization by gyroscopic effects is possible [47]. Experiments revealed extreme stiffness was maintained over many cycles of deformation; the stability of these composites is not fully understood. It is possible that energy stored during a constrained phase transformation is released gradually, giving rise to energy flux (§2.5.5).

As for waves, stable composites with negative stiffness components [48] have been shown give rise to extremely high values of wave attenuation (§9.3.5) and can block waves over an extended range of frequency.

Biological tissue negative properties

Negative stiffness has been inferred for active cells in the ear which are supplied with a metabolic power source [49] [50]. The active cells, called stereocilia, are reduced in stiffness even to negative stiffness via a gating stimulus. Cells in the inner ear also perform amplification [51], so the mechanical damping is negative. The active amplification of these cells can give rise to pure tones emitted from the ear, spontaneously or in response to acoustic stimulation [52].

2.5.5 Negative and extreme moduli: energy flux

In contrast to the systems above which contain stored energy, one may consider systems subject to energy flux. This is clearly a non-equilibrium condition. Negative spring constants can be stabilized in articulated groups of pipes [54] carrying moving fluid. These pipe systems are not in equilibrium; a source of power is supplied to move the fluid. The flow rate can be adjusted to attain negative stiffness or extremely high structural stiffness that is stable. Stiffness of coupled field systems is tunable from negative stiffness to large positive stiffness by a varying temperature difference that drives a flow of heat or a voltage that drives a flow of electric current [55]. These systems evade thermodynamic restrictions by relaxing a restrictive assumption: equilibrium.

2.5.6 Negative heat capacity

As for heat capacity, heat capacity is negative in stars and star clusters [56] [57] in which there is a long range gravitational interaction. The contradiction with thermodynamic "proofs" is explained [57] by observing that an extensive system (in which a canonical ensemble of sub-systems is in equilibrium) is implicitly assumed in thermodynamic analyses. Stars are not extensive systems (with many identical subsystems) and they are not in equilibrium. The assumptions underlying the proof of positive heat capacity do not apply. Negative heat capacity can occur in materials, such as [58] rapidly cooled amorphous materials near the glass transition temperature, that are not in equilibrium. Negative heat capacity also occurs in the liquid-to-gas transition of finite systems such as clusters of ions consisting of 139 atoms [59].

2.5.7 Negative capacitance

The electrical capacitance quantifies the ability of a circuit element to store electric charge. The energy is given by $U = \frac{1}{2}CV^2$ in which C is the capacitance and V is the applied voltage. As with deformation energy, positive definiteness of the electrical energy implies the capacitance is positive. The electrical system, however, may be provided with an initial energy or with a power source, allowing negative capacitance. Negative capacitance in feedback amplifiers has long been known. Such amplifiers have been used [60] to neutralize some of the positive capacitance of transducers in order to improve their damping properties. Sound isolation by piezoelectric polymer films was enhanced by connection to negative capacitance circuits [61]. Negative incremental capacitance was predicted [62] to occur in ferroelectric materials via the classic Landau theory of phase transformations. More recently negative capacitance was inferred in a ferroelectric field effect transistor [63] and in ferroelectrics [64]. Possible applications have been envisaged in reducing power use in micro-circuits.

2.6 Summary

Among inclusion shapes in dilute randomly oriented two-phase composites, stiff spheres provide the least stiffening effect; stiff platelets provide the most; fibers are intermediate. Spherical inclusions provide a modest stiffening effect even if they are much stiffer than the matrix; in the limit for $G_2 \to \infty$ and for matrix Poisson's ratio of $1/2$, inclusion stiffness $\frac{G_c}{G_1} \approx 1 + 2.5V_2$. Fiber or platelet inclusions of high stiffness, by contrast, result in a proportionally stiff composite. The greatest stiffness is obtained when fibers or platelets are oriented in the direction of the stress. If stress can come from any direction, then it is sensible to orient the inclusions randomly. The stiffness of such a composite is considerably less than that of an oriented one, particularly for fibrous materials.

Extremal stiffness attaining the bounds can be achieved via hierarchical microstructures including coated spheres of different size, and hierarchical laminates with a range of lamina thicknesses. Hierarchical laminates have also been used to attain bounds on other physical properties such as conductivity.

Bounds can be exceeded by relaxing assumptions used to derive them. For example, a material need not be in a minimum energy state or be capable of infinite subdivision; it need not be in an environment of zero energy flux. Stored energy corresponding to a non-minimum energy state may be incorporated by applying a pre-strain or in the context of phase transformations. Extremely high values of stiffness or of viscoelastic damping can be attained by such approaches.

Bibliography

[1] R. S. Lakes, *Viscoelastic Materials*, Cambridge University Press, Cambridge (2009).

[2] G. W. Milton, *The Theory of Composites*, Cambridge University Press, Cambridge (2002).

[3] I. S. Sokolnikoff, *Theory of Elasticity*, Krieger; Malabar, FL, (1983).

[4] S. P. Timoshenko and J. N. Goodier, *Theory of Elasticity*, McGraw Hill, New York (1982).
[5] D. C. Wallace, *Thermodynamics of crystals.* J. Wiley, New York (1972).
[6] R. S. Lakes, Foam structures with a negative Poisson's ratio, Science, 235, 1038-1040, (1987).
[7] G. Milton, Composite Materials with Poisson's Ratios Close to -1, J. Mech. Phys. Solids, 40, 1105-1137 (1992).
[8] Y. C. Wang and R. S. Lakes, Composites with inclusions of negative bulk modulus: extreme damping and negative Poisson's ratio, J. Composite Materials, 39, 1645-1657, (2005).
[9] E. F. Cosserat, Sur les equations de la théorie de l'elasticité, C. R. Acad. Sci. Paris 126, 1089-1091 (1898).
[10] J. K. Knowles, and E.Sternberg, On the failure of ellipticity and the emergence of discontinuous gradients in plane finite elastostatics, J. Elasticity, 8, 329-379, (1978).
[11] R. Hill, The Elastic behaviour of a crystalline aggregate, Proc. Phys. Soc. A 65, 349-354, (1952).
[12] R. Hill, Elastic properties of reinforced solids: Some theoretical principles, Journal of the Mechanics and Physics of Solids 11, 357-372, (1963).
[13] Z. Hashin, The elastic moduli of heterogeneous materials, J. Appl. Mech., Trans. ASME, 84E, 143-150, (1962).
[14] Z. Hashin and S. Shtrikman, A variational approach to the theory of the elastic behavior of multiphase materials, J. Mech. Phys. Solids, 11, 127-140, (1963).
[15] Z. Hashin, Analysis of composite materials- a survey, J. Applied Mechanics, 50, 481-505, (1983).
[16] K. A. Lurie and A. V. Cherkaev, Exact estimates of the conductivity of a binary mixture of isotropic materials, Proceedings of the Royal Society of Edinburgh, 104A, 21-38, (1986).
[17] G. W. Milton, Modeling the properties of composites by laminates. in Erickson, J. L., Kinderlehrer, D., Kohn, R., Lions, J. L. (Eds.), *Homogenization and Effective Moduli of Materials and Media.* Springer Verlag, Berlin, 150-175, (1986).
[18] G. A. Francfort and F. Murat, Homogenization and optimal bounds in linear elasticity, Arch. Rational Mech. Anal. 94(4), 307-334 (1986).
[19] G. W. Milton and S. K. Serkov, Neutral coated inclusions in conductivity and anti-plane elasticity. Proceedings of the Royal Society 457, 1973-1997, (2001).
[20] L. V. Gibiansky and O. Sigmund, Multiphase composites with extremal bulk modulus, Journal of the Mechanics and Physics of Solids, 48(3), 461-498 (2000).

[21] J. C. Maxwell, *A Treatise on Electricity and Magnetism*, Article 322, p. 371-372, Oxford, Clarendon Press, Oxford, U.K. (1873).
[22] K. Schulgasser, Bounds on the conductivity of statistically isotropic polycrystals, Journal of Physics C: Solid State Physics, 10, 407-417 (1977).
[23] M. Avellaneda, Optimal bounds and microgeometries for elastic two-phase composites, SIAM Journal on Applied Mathematics, 47(6), 1216-1228 (1987).
[24] G. W. Milton, A.V. Cherkaev, Which elasticity tensors are realizable? ASME J. Eng. Mater. Technol. 117, 483-493 (1995).
[25] R. M. Christensen, Viscoelastic properties of heterogeneous media, J. Mech. Phys. Solids, 17, 23-41, (1969).
[26] H. J. Kim, C. C. Swan and R. S. Lakes, Computational studies on high stiffness, high damping SiC-InSn particulate reinforced composites, Int. J. Solids, Structures, 39, 5799-5812, (2002).
[27] R. M. Christensen, *Mechanics of Composite Materials*, John Wiley & Sons, New York, (1979).
[28] R. M. Christensen, Isotropic properties of platelet reinforced media, J. Eng. Mat. Tech. 101, 299-303, (1979).
[29] A. N. Norris, The mechanical properties of platelet reinforced composites, Int. J. Solids, Structures, 26, 663-674, (1990).
[30] R. Roscoe, Isotropic composite with elastic or viscoelastic phases: general bounds for the moduli and solutions for special geometries, Rheologica Acta, 12, 404-411, (1973).
[31] Z. Bazant and L. Cedolin, *Stability of Structures*, Oxford University Press, Oxford, (1991).
[32] U. S. Patent 5310157, Vibration isolation system, Minus K Technology, Inc. (1994).
[33] R. S. Lakes, Extreme damping in compliant composites with a negative stiffness phase, Philosophical Magazine Letters, 81, 95-100 (2001).
[34] L. Dong and R. S. Lakes, Advanced damper with high stiffness and high hysteresis damping based on negative structural stiffness, Int. J. Solids Struct., 50, 2413-2423 (2013).
[35] S. Balch, and R. S. Lakes, Lumped negative stiffness damper for absorption of flexural waves in a rod, Smart Materials and Structures, 26 045022 (6pp) (2017).
[36] S. Leelavanichkul, A. V. Cherkaev, D. O. Adams, F. Solzbacher, Energy absorption of a helicoidal bistable structure, Journal of Mechanics of Materials and Structures 5(2), 305-321 (2010).
[37] R. E. McKnight, A. T. Moxon, A. Buckley, P. A. Taylor, T. W. Darling, and M. A. Carpenter, Grain size dependence of elastic anomalies accompanying the alpha-beta phase transition in polycrystalline quartz. J. Phys. Cond. Mat. 20, 075229 (2008).

[38] S. Hirotsu, Elastic anomaly near the critical point of volume phase transition in polymer gels, Macromolecules, 23, 903-905, (1990).

[39] L. Dong, D. S. Stone, R. S. Lakes, Softening of bulk modulus and negative Poisson's ratio in barium titanate ceramic near the Curie point. Philos. Mag. Lett. 90, 23-33, (2010).

[40] T. Jaglinski, D. Kochmann, D. Stone, R. S. Lakes, Materials with viscoelastic stiffness greater than diamond, Science 315, 620-622, Feb. 2 (2007).

[41] R. S. Lakes, Extreme damping in composite materials with a negative stiffness phase. Phys. Rev. Lett. 86, 2897-2900 (2001).

[42] R. S. Lakes, T. Lee, A. Bersie, and Y. C. Wang, Extreme damping in composite materials with negative stiffness inclusions, Nature, 410, 565-567, (2001).

[43] R. S. Lakes, and W. J. Drugan, Dramatically stiffer elastic composite materials due to a negative stiffness phase?, J. Mechanics and Physics of Solids, 50, 979-1009 (2002).

[44] X. Shang and R. S. Lakes, Stability of elastic material with negative stiffness and negative Poisson's ratio, Physica Status Solidi (b), 244, 1008-1026, (2007).

[45] W. J. Drugan, Elastic composite materials having a negative stiffness can be stable, Phys. Rev. Letters, 98, 055502, (2007).

[46] C. S. Wojnar and D. M. Kochmann, A negative-stiffness phase in elastic composites can produce stable extreme effective dynamic but not static stiffness, Philos. Mag. 94 532-555 (2014).

[47] D. M. Kochmann and W. J. Drugan, An infinitely-stiff elastic system via a tuned negative-stiffness component stabilized by rotation-produced gyroscopic forces, Appl. Phys. Lett. 108, 261904 (2016).

[48] W. J. Drugan, Wave propagation effects possible in solid composite materials by use of stabilized negative-stiffness components, J. Mechanics and Physics of Solids, `https://doi.org/10.1016/j.jmps.2019.103700`

[49] P. Martin, A. D. Mehta, and A. J. Hudspeth, Negative hair-bundle stiffness betrays a mechanism for mechanical amplification by the hair cell, Proc. Natl. Acad. Sci. U.S.A. 97, 12026-12031 (2000).

[50] K. H. Iwasa, G. Ehrenstein, Cooperative interaction as the physical basis of the negative stiffness in hair cell stereocilia, The Journal of the Acoustical Society of America, 111 2208 (2002).

[51] T. Gold, Hearing. II. The physical basis of the action of the cochlea, Proceedings of the Royal Society of London. Series B, Biological Sciences, 135, 492-498, (1948).

[52] D. T. Kemp, Stimulated acoustic emissions from within the human auditory system, J. Acoust. Soc. Am., 64, 1386-1391, (1978).

[53] L. D. Landau, in *Collected papers of L. D. Landau*, ed. D. Ter Taar, Gordon and Breach / Pergamon, New York, London (1965).
[54] J. M. T. Thompson, 'Paradoxical' mechanics under fluid flow. Nature. 296, 135-137 (1982).
[55] R. S. Lakes, Stable singular or negative stiffness systems in the presence of energy flux, Philosophical Magazine Letters, 92, 226-234, (2012).
[56] D. Lynden Bell and R. Wood, The gravo-thermal catastrophe in isothermal spheres and the onset of redgiant structure in stellar systems. Monthly Notices of the Royal Astronomical Society 138, 495-525, (1968).
[57] D. Lynden Bell, Negative specific heat in astronomy, physics, and chemistry. Physica A, 263, 293-304 (1999).
[58] J. Bisquert, Master equation approach to the non-equilibrium negative specific heat at the glass transition, Am. J. Phys. 73, 735-741 (2005).
[59] F. Gobet, B. Farizon, M. Farizon, M. J. Gaillard, J. P. Buchet and M. Carré, P. Scheier and T. D. Märk, Direct experimental Evidence for a negative heat capacity in the liquid-to-gas phase transition in hydrogen cluster Ions: backbending of the caloric curve, Phys. Rev. Lett. 89(18), 183403 (2002).
[60] R. L. Forward, U. S. Patent number: 4158787, Electromechanical transducer-coupled mechanical structure with negative capacitance compensation circuit, (1979).
[61] E. Fukada, M. Date, K. Kimura, T. Ohkubo, H. Kodama, P. Mokry and K. Yamamoto, Sound isolation by piezoelectric polymer films connected to negative capacitance circuits, IEEE Transactions on dielectrics and electrical insulation, 11(2) 328-332 (2004).
[62] L. D. Landau and I. M. Khalatnikov, On the anomalous absorption of sound near a second order phase transition point, Dokl. Akad. Nauk 96, 469-772 (1954).
[63] G. A. Salvatore, A. Rusu, and A. M. Ionescu, Experimental confirmation of temperature dependent negative capacitance in ferroelectric field effect transistor, Applied Physics Letters 100, 163504 (2012).
[64] D. Appleby, N. K. Ponon, K. Kwa, B. Zou, P. K. Petrov, T. Wang, N. M. Alford, and A. O'Neill, Experimental observation of negative capacitance in ferroelectrics at room temperature, Nano Lett., 14, 3864-3868 (2014).

Chapter 3

Symmetry and anisotropy

3.1 Introduction and Rationale

Anisotropy is associated with the symmetry of the structure within a material. The structure may be on the atomic scale as in single crystals or it may be on a larger scale, even a macroscopic scale as in honeycombs, foams, and truss (rib) lattices.

The anisotropic elastic behavior of materials is pertinent to the deformation of single crystals and of composite materials that have preferred structural orientation. Such materials may behave very differently depending on the direction of the applied load. Anisotropic stress-strain relations are also used in the design of laminates made of anisotropic layers (laminae) containing oriented fibers. Anisotropic composites offer superior strength and stiffness in comparison with isotropic ones. Material properties in one direction are gained at the expense of properties in other directions. It is sensible, therefore, to use anisotropic composite materials only if the direction of application of the stress is known in advance.

Symmetry and anisotropy pertain to other physical properties such as thermal expansion, piezoelectricity, and pyroelectricity which entail coupled fields. Properties associated with coupled fields are examined in Chapter 4. Some physical properties such as piezoelectricity and pyroelectricity require a degree of asymmetry in the material structure in order to be manifest.

3.2 Tensors

Tensors are defined in the context of transformation under a change of coordinates. A scalar is a single number that remains the same under a change of coordinates. It is a tensor of rank zero.

A vector is, in three dimensions, a set of three numbers that transforms as $v'_i = a_{ij} v_j$. Indices i and j can assume values 1,2,3 in three dimensions. Repeated indices are assumed to be summed over: the Einstein summation convention is discussed in detail in the next section. The primed coordinate system is generated by the transformation matrix a_{ij}. The transformation may be a rotation of coordinates. Velocity is an example of a vector. A vector is a tensor of rank one.

A tensor of rank two transforms as $\sigma'_{ij} = a_{im} a_{jn} \sigma_{mn}$. There are nine numbers comprising the tensor σ_{ij} which could represent a stress. Tensors of higher rank are defined similarly by increasing the number of indices.

Physical properties can be represented by tensors. For example, the heat capacity is a tensor of rank zero; the thermal expansion and the conductivity are tensors of rank two. The piezoelectric sensitivity is a tensor of rank three, and the elastic modulus is a tensor of rank four.

Higher rank entails a greater amount of freedom in the physical properties. For example, the thermal expansion of an isotropic material reduces to a tensor with only diagonal elements that are all equal, so we think of the thermal expansion of isotropic materials as a number. In isotropic elastic materials, there are two independent elastic constants. Also, there is a distinction between even and odd rank tensors. Effects of chirality in a material (it has left and right handed forms) are manifest only in tensor properties of odd rank.

3.3 Elastic properties

3.3.1 Hooke's law

Composites may be isotropic or anisotropic. Properties of anisotropic materials depend on direction; isotropic materials have no preferred direction. Composites containing spherical particulate inclusions distributed randomly in an isotropic matrix are macroscopically isotropic. Unidirectional fibrous composites and laminates are highly anisotropic if the phases differ greatly in modulus. Too, single crystals are anisotropic. Laminates containing fibrous layers can be prepared with a controlled degree of anisotropy. Anisotropy can be dealt with using a tensorial stress-strain relation. For elastic materials, Hooke's law of linear elasticity is given in the modulus formulation by

$$\sigma_{ij} = C_{ijkl} \epsilon_{kl}, \tag{3.1}$$

with C_{ijkl} as the elastic modulus tensor, ϵ_{kl} as strain, σ_{ij} as stress, and the usual Einstein summation convention assumed in which repeated indices are summed over [1] [2] [4]. More explicitly,

$$\sigma_{ij} = \sum_{k=1}^{3} \sum_{l=1}^{3} C_{ijkl}\epsilon_{kl}. \tag{3.2}$$

The first of the nine equations represented here begins as follows.

$$\sigma_{11} = C_{1111}\epsilon_{11} + C_{1112}\epsilon_{12} + C_{1113}\epsilon_{13} + C_{1122}\epsilon_{22} + C_{1123}\epsilon_{23} + \cdots \tag{3.3}$$

There are 81 components of C_{ijkl} (or of the compliance J_{ijkl}), but taking into account the symmetry of the stress and strain tensors, only 36 of them are independent. Recall that the strain is symmetric by its definition. The symmetry of the stress is shown by an equilibrium argument. An asymmetric stress would generate an unbalanced torque on a differential element, hence divergent angular acceleration. That is true provided there are no other torques such as distributed magnetic torques in magnetic materials, or distributed torque per area as in materials with nontrivial structure size (Chapter 8).

If the elastic solid is describable by a strain energy function, the number of independent elastic constants is reduced to 21 by virtue of the resulting symmetry $C_{ijkl} = C_{klij}$. An elastic modulus tensor with 21 independent constants describes an anisotropic material with the most general type of anisotropy, triclinic symmetry.

In the compliance formulation,

$$\epsilon_{ij} = J_{ijkl}\sigma_{kl} \tag{3.4}$$

with J_{ijkl} as the compliance tensor. Some authors denote the compliance by S or s rather than J.

3.3.2 Reduced notation: matrix form

The modulus and compliance tensors are at times written [6] [7] [5] in a six by six matrix form by using the correspondence $11 \rightarrow 1, 22 \rightarrow 2, 33 \rightarrow 3, 23 \rightarrow 4, 13 \rightarrow 5, 12 \rightarrow 6$. In this reduced notation, $C_{1111} \rightarrow C_{11}$ and $C_{2323} \rightarrow C_{44}$. The reduced notation is used because a modulus representation with four subscripts represents a hyper-cubical array in four dimensional space, difficult to visualize. The modulus does not transform (under coordinate changes) as a planar six by six matrix; it still transforms as a fourth rank tensor. The reduced notation allows the modulus to be presented, in a two-dimensional format such as on paper, as a matrix.

Hooke's law in matrix form is written $\sigma = \mathbf{C}\epsilon$. The strain ϵ is written as a column vector with the shear strain taken as engineering shear, a factor of two larger than tensorial shear strain. The modulus matrix $\mathbf{C}$ is not a tensor and does not transform as one.

3.3.3 Symmetry classes

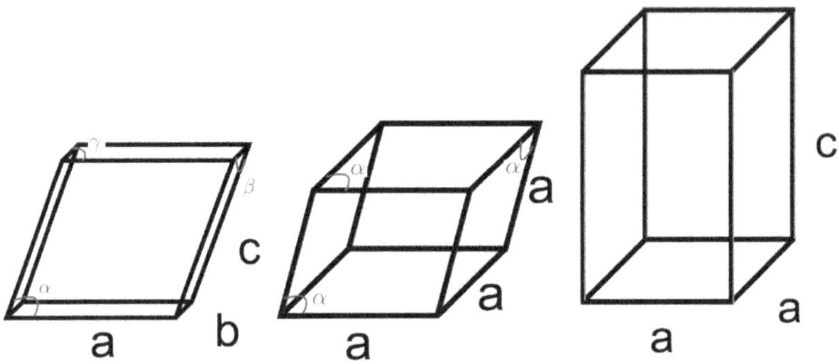

Figure 3.1: Unit cell shapes for triclinic, rhombohedral, and tetragonal symmetry. For triclinic symmetry, side lengths are unequal; angles are unequal: $\alpha \neq \beta \neq \gamma$. For rhombohedral symmetry, side lengths are equal; angles are oblique and equal. For tetragonal symmetry, one side length is unequal, $a = a \neq c$; all angles are right angles.

The symmetry classes triclinic, rhombohedral, tetragonal, orthorhombic, hexagonal and cubic, shown in Figures 3.1 and 3.2, are most familiar for single crystals in which the structure is on the scale of atoms. The symmetry concepts also apply to materials in which the structure size is on a larger scale, from nano to micro to macroscopic. Unit cell shapes are considered to repeat in three dimensions in the material.

For a *triclinic* crystal or material, the unit cell has three different side lengths $a \neq b \neq c$ and three different angles, $\alpha \neq \beta \neq \gamma$, shown in Figure 3.1. For a *monoclinic* crystal or material, the unit cell has three different side lengths $a \neq b \neq c$; two axes are inclined at a 90° angle; the third axis is oblique.

For a *tetragonal* crystal or material, the unit cell has all right angles and two side lengths equal and one different. For an *orthorhombic* crystal or *orthotropic* material, the unit cell shown in Figure 3.2 has three unequal sides and all right angles. There are three orthogonal planes of symmetry. The *hexagonal* structure is invariant to rotations of 60° angle. For a *cubic* crystal or material, the unit cell shown in Figure 3.2 has three equal sides and all right angles. The cubic structure is the most symmetric of the crystal classes. *Isotropic* materials have microstructure that is indistinguishable in all directions and they have the same physical properties in all directions.

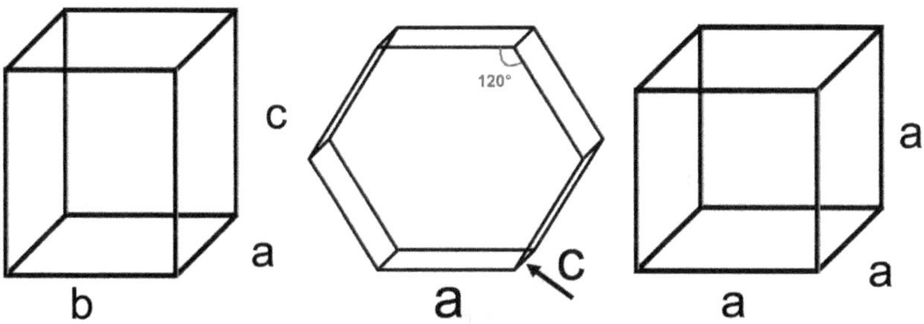

Figure 3.2: Unit cell shapes for orthorhombic, hexagonal, and cubic symmetry. For orthorhombic symmetry, side lengths are unequal, $a \neq b \neq c$; all angles are right angles. For hexagonal symmetry, the internal angle is $\gamma = 120°$. For cubic symmetry, side lengths are equal; all angles are right angles.

3.3.4 Quasicrystals

In standard crystallography it can be proven that only certain shapes can be packed without gaps to fill a planar region or a three dimensional space. Fivefold symmetry applies to a single regular pentagon but such pentagons cannot be packed to fill a plane unless there are gaps. Similarly, icosahedra cannot be packed to fill a three dimensional space without gaps. However a rapidly cooled Al 14 atomic % Mn alloy exhibits long-range orientational order as revealed by diffraction patterns [8] as discovered by Schechtman, who won the Nobel Prize. The material has icosahedral point group symmetry. Such symmetry is inconsistent with unit cells that exhibit lattice translations. Materials of this type are called quasicrystals. The microstructure exhibits rotational symmetry that is revealed in diffraction patterns but the translation symmetry is not fully periodic.

Penrose tiling patterns [9] [10] [11] represent a two dimensional counterpart to a quasicrystal. As with the quasicrystal, there is rotational order but in translation the pattern is not periodic. A Penrose tiling may be made with two quadrilateral figures called kites and darts. The kite is a convex quadrilateral which has four interior angles 72, 72, 72, and 144 degrees. The dart is a non-convex quadrilateral which has four interior angles 36, 72, 36, and 216 degrees.

3.3.5 Modulus matrices and symmetry

For a triclinic material, the elastic modulus matrix is fully populated. There are 36 elastic constants. In an elastic material describable by a strain energy density [6], only 21 of them are independent. The strain energy density is $U = \frac{1}{2}C_{ijkl}\epsilon_{ij}\epsilon_{kl}$. Interchange of ij and kl makes no difference so $C_{ijkl} = C_{klij}$ so in the reduced notation, $C_{ij} = C_{ji}$. The matrix in the reduced notation is symmetric.

$$\mathbf{C} = \begin{pmatrix} C_{11} & C_{12} & C_{13} & C_{14} & C_{15} & C_{16} \\ C_{21} & C_{22} & C_{23} & C_{24} & C_{25} & C_{26} \\ C_{31} & C_{32} & C_{33} & C_{34} & C_{35} & C_{36} \\ C_{41} & C_{42} & C_{43} & C_{44} & C_{45} & C_{46} \\ C_{51} & C_{52} & C_{53} & C_{54} & C_{55} & C_{56} \\ C_{61} & C_{62} & C_{63} & C_{64} & C_{65} & C_{66} \end{pmatrix} \tag{3.5}$$

In a monoclinic solid, in comparison with a triclinic solid, some of the cross coupling constants vanish but some remain. Equation 3.6 is for invariance parallel to the x_2 direction: $x_1 \to -x_1$, $x_2 \to x_2$, $x_3 \to -x_3$. If a different invariance direction is chosen, the nonzero coupling terms will appear in different locations. *Monoclinic* materials have 13 independent elastic constants.

$$\mathbf{C} = \begin{pmatrix} C_{11} & C_{12} & C_{13} & 0 & C_{15} & 0 \\ C_{21} & C_{22} & C_{23} & 0 & C_{25} & 0 \\ C_{31} & C_{32} & C_{33} & 0 & C_{35} & 0 \\ 0 & 0 & 0 & C_{44} & 0 & C_{46} \\ C_{51} & C_{52} & C_{53} & 0 & C_{55} & 0 \\ 0 & 0 & 0 & C_{64} & 0 & C_{66} \end{pmatrix} \tag{3.6}$$

The modulus matrix of an orthorhombic or orthotropic material has nine independent elastic constants. This matrix is symmetric, as with all the modulus matrices here, for elastic materials for which there is a strain energy function so that $C_{ij} = C_{ji}$. For example, $C_{13} = C_{31}$.

$$\mathbf{C} = \begin{pmatrix} C_{11} & C_{12} & C_{13} & 0 & 0 & 0 \\ C_{21} & C_{22} & C_{23} & 0 & 0 & 0 \\ C_{31} & C_{32} & C_{33} & 0 & 0 & 0 \\ 0 & 0 & 0 & C_{44} & 0 & 0 \\ 0 & 0 & 0 & 0 & C_{55} & 0 \\ 0 & 0 & 0 & 0 & 0 & C_{66} \end{pmatrix} \tag{3.7}$$

For an *hexagonal* crystal or an *axisymmetric* material, the modulus matrix has five independent elastic constants. Some pairs of moduli are equal as shown in the matrix, Equation 3.8.

$$\mathbf{C} = \begin{pmatrix} C_{11} & C_{12} & C_{13} & 0 & 0 & 0 \\ C_{21} & C_{11} & C_{13} & 0 & 0 & 0 \\ C_{31} & C_{31} & C_{33} & 0 & 0 & 0 \\ 0 & 0 & 0 & C_{44} & 0 & 0 \\ 0 & 0 & 0 & 0 & C_{44} & 0 \\ 0 & 0 & 0 & 0 & 0 & \frac{C_{11}-C_{12}}{2} \end{pmatrix} \tag{3.8}$$

The modulus matrix of a *cubic* material has three independent elastic constants. Some elastic constants are equal to others as shown in Equation 3.9. The cubic structure is invariant to rotations of a 90° angle about any of the three coordinate directions.

$$\mathbf{C} = \begin{pmatrix} C_{11} & C_{12} & C_{12} & 0 & 0 & 0 \\ C_{12} & C_{11} & C_{12} & 0 & 0 & 0 \\ C_{12} & C_{12} & C_{11} & 0 & 0 & 0 \\ 0 & 0 & 0 & C_{44} & 0 & 0 \\ 0 & 0 & 0 & 0 & C_{44} & 0 \\ 0 & 0 & 0 & 0 & 0 & C_{44} \end{pmatrix} \tag{3.9}$$

As for other symmetry classes, *tetragonal* materials have 6 or 7 depending on the specific class, and *trigonal* (*rhombohedral*) materials have 6 or 7 depending on the specific class.

The symmetry arguments apply equally well to the compliance formulation; the number of elastic constants for each symmetry does not change.

Example: elastic constants under rotation

Consider rotation of a specimen through a 90° angle, as indicated by the following rotation matrix.

$$\mathbf{a} = \begin{pmatrix} 1 & 0 & 0 \\ 0 & 0 & 1 \\ 0 & -1 & 0 \end{pmatrix} \tag{3.10}$$

How does the rotation affect the elastic constants, specifically C_{2222}?

Recall $C'_{ijkl} = a_{im}a_{jn}a_{ko}a_{lp}C_{mnop}$, with an implied sum on repeated indices. So $C'_{2222} = a_{2m}a_{2n}a_{2o}a_{2p}C_{mnop}$. This is still an implied sum. The only element of the rotation matrix **a** that has 2 as its first index is $a_{23} = 1$. So $C'_{2222} = a_{23}a_{23}a_{23}a_{23}C_{3333} = C_{3333}$.

If the material is invariant to such rotations then $C'_{2222} = C_{2222}$.

So $C_{2222} = C_{3333}$. If the material is invariant to all 90° rotations about three mutually orthogonal axes then it has cubic symmetry. Analyses similar to the above disclose for cubic materials all cross coupling terms such as stretch-shear vanish, the three axial constants are equal, and the three shear moduli are equal; there are three independent elastic constants.

3.3.6 Isotropy

Elastic constants

For an *isotropic* material, $\frac{1}{2}(C_{11} - C_{12}) = C_{44}$ so there are two independent elastic constants. Four constants are commonly used, Young's modulus E, shear modulus G, bulk modulus K (inverse compressibility) and Poisson's ratio ν; these are interrelated. The bulk modulus is at times called B. For example, the bulk modulus is $K = 2G\frac{1+\nu}{3(1-2\nu)}$, and $E = 2G(1+\nu)$. Further conditions on the elastic constants are imposed in the context of stability. To be stable, a material with free surfaces must be in a minimum energy state at zero strain. For an isotropic object with free surfaces to be stable, $G > 0$, $K > 0$, which give, in three dimensions, $-1 < \nu < \frac{1}{2}$.

Modulus matrix for isotropic solid

In an isotropic solid, we have $C_{1111} = E\frac{1-\nu}{(1+\nu)(1-2\nu)}$. This is C_{11} in the reduced notation. This may be demonstrated either using the tensor form as shown in the following or using elementary analysis in the compliance formulation as shown in the subsequent subsection.

Relate the modulus C_{1111} to Young's modulus E and Poisson's ratio using the modulus tensor formulation.

Proceed first to consider elastic materials via the tensorial version of Hooke's law, then the elementary three-dimensional form. In three dimensions, Hooke's law of linear elasticity is given in tensorial form by: $\sigma_{ij} = C_{ijkl}\epsilon_{kl}$. The index notation, with i, j, k, and l each able to assume values from 1 to 3 is used as is the Einstein summation convention in which repeated indices are summed over.

Consider simple tension or compression in the 1 (or x) direction: $\sigma_{11} \neq 0$, $\sigma_{22} = 0$, $\sigma_{11} = 0$. Expand the sum, assuming orthotropic symmetry.

$$\sigma_{11} = C_{1111}\epsilon_{11} + C_{1122}\epsilon_{22} + C_{1133}\epsilon_{33}.$$

Use the definition of Poisson's ratio to write this in mixed tensorial and engineering symbols.

$$\sigma_{11} = C_{1111}\epsilon_{11} + C_{1122}(-\nu_{12}\epsilon_{11}) + C_{1133}(-\nu_{13}\epsilon_{11}).$$

The ratio of stress to strain in simple tension in the 1 direction is Young's modulus E_1 in the 1 direction.

$$E_1 = \frac{\sigma_{11}}{\epsilon_{11}} = C_{1111} - \nu_{12}C_{1122} - \nu_{13}C_{1133}.$$

Consider the 2 (or y) direction in which there is zero stress, and again incorporate the definition of the Poisson's ratios.

$$0 = \sigma_{22} = C_{2211}\epsilon_{11} + C_{2222}\epsilon_{22} + C_{2233}\epsilon_{33} = \{C_{2211} - \nu_{12}C_{2222} - \nu_{13}C_{2233}\}\epsilon_{11}.$$

Assume the material is isotropic, so both the stiffness and the Poisson's ratio are independent of direction and $C_{1111} = C_{2222}$. Then, $C_{2211} - \nu C_{2211} = \nu C_{1111}$; or $C_{2211} = \frac{\nu}{1-\nu}C_{1111}$.

Substituting above, and dropping the subscript on E since in an isotropic material there is no dependence of properties on direction,

$$E = [1 - \tfrac{2\nu^2}{1-\nu}]C_{1111}$$

so

$$C_{1111} = E\frac{1-\nu}{(1+\nu)(1-2\nu)}. \tag{3.11}$$

Since $E = 2G(1+\nu)$ for isotropic materials and $\nu = \frac{3K-2G}{6K+2G}$, the constrained modulus tensor element may be written in terms of the bulk modulus K

$$C_{1111} = K + \tfrac{4}{3}G.$$

Consider C_{12}.

To determine C_{12}, consider uniaxial stress in the x direction with other stresses equal to zero.

$$\sigma_{22} = 0 = C_{2211}\epsilon_{11} + C_{2222}\epsilon_{22} + C_{2233}\epsilon_{33}$$

Use the definition of Poisson's ratio $\nu = -\frac{\epsilon_{22}}{\epsilon_{11}} = -\frac{\epsilon_{33}}{\epsilon_{11}}$.

The material is isotropic so the Poisson's ratios are equal and $C_{2222} = C_{1111}$; in the reduced notation this is C_{11}.

So

$$0 = (C_{2211} - \nu C_{2222} - \nu C_{2233})\epsilon_{11}$$

Also $C_{2211} = C_{2233}$. In the reduced notation, $C_{21} = C_{12} = C_{23}$ so

$$C_{12} = C_{11}\tfrac{\nu}{1-\nu} = \tfrac{E\nu}{(1+\nu)(1-2\nu)}.$$

Recall the shear moduli are all equal and that $C_{44} = G$.

Finally, Eq. 3.12 shows the elastic modulus matrix in terms of the engineering elastic constants.

$$\mathbf{C} = \begin{pmatrix} \frac{E(1-\nu)}{(1+\nu)(1-2\nu)} & \frac{E\nu}{(1+\nu)(1-2\nu)} & \frac{E\nu}{(1+\nu)(1-2\nu)} & 0 & 0 & 0 \\ \frac{E\nu}{(1+\nu)(1-2\nu)} & \frac{E(1-\nu)}{(1+\nu)(1-2\nu)} & \frac{E\nu}{(1+\nu)(1-2\nu)} & 0 & 0 & 0 \\ \frac{E\nu}{(1+\nu)(1-2\nu)} & \frac{E\nu}{(1+\nu)(1-2\nu)} & \frac{E(1-\nu)}{(1+\nu)(1-2\nu)} & 0 & 0 & 0 \\ 0 & 0 & 0 & G & 0 & 0 \\ 0 & 0 & 0 & 0 & G & 0 \\ 0 & 0 & 0 & 0 & 0 & G \end{pmatrix} \tag{3.12}$$

The stress-strain relation for isotropic materials is at times written

$$\sigma_{ij} = 2G\epsilon_{ij} + \lambda\epsilon_{kk}\delta_{ij} \tag{3.13}$$

in which G is the shear modulus and λ is a second elastic constant; the usual Einstein summation convention for repeated indices is used in which repeated indices are summed over.

Modulus C_{1111} via elementary method

Determine the physical meaning of the modulus C_{1111} for isotropic materials. Specifically, relate this modulus to Young's modulus and Poisson's ratio. Use the elementary isotropic formulation of Hooke's law in three dimensions. This is in the compliance formulation because $1/E = J$.

Solution

Write the elementary isotropic form for Hooke's law. This is an expanded form of the compliance formulation.

$$\epsilon_{xx} = \frac{1}{E}\{\sigma_{xx} - \nu\sigma_{yy} - \nu\sigma_{zz}\} \tag{3.14}$$

$$\epsilon_{yy} = \frac{1}{E}\{\sigma_{yy} - \nu\sigma_{xx} - \nu\sigma_{zz}\} \tag{3.15}$$

$$\epsilon_{zz} = \frac{1}{E}\{\sigma_{zz} - \nu\sigma_{xx} - \nu\sigma_{yy}\} \tag{3.16}$$

Consider simple tension or compression in the x direction: $\sigma_{xx} \neq 0$, $\sigma_{yy} = 0$, $\sigma_{zz} = 0$. Then

$$\frac{\sigma_{xx}}{\epsilon_{xx}} = E.$$

Consider constrained compression, with $\epsilon_{yy} = 0$, $\epsilon_{zz} = 0$. Then

$$\sigma_{yy} = \nu\sigma_{xx} + \nu\sigma_{zz}.$$
$$\sigma_{zz} = \nu\sigma_{xx} + \nu\sigma_{yy}.$$

Substituting,

$$\sigma_{yy} = \sigma_{zz} = \sigma_{xx}\frac{\nu(1+\nu)}{1-\nu^2}.$$

So, substituting into Hooke's law, the stress-strain ratio for constrained compression, which by definition is the constrained modulus C_{1111}, is

$$\frac{\sigma_{xx}}{\epsilon_{xx}} = C_{1111} = E\frac{1-\nu}{(1+\nu)(1-2\nu)}.$$

3.3.7 Physical interpretation: elastic modulus

In the *modulus formulation*, the constant C_{1111} is a modulus representing the ratio of stress to strain in the 1 or x direction, with all other *strain* components zero. It is revealed by a confined compression test or an ultrasonic test at high frequency for which the Poisson strain is constrained to be zero. C_{1111} is not equal to a Young's modulus unless Poisson's ratio is zero. As for the interpretation of C_{1111} for isotropic elastic materials we note that [1] [2],

$$C_{1111} = E\frac{1-\nu}{(1+\nu)(1-2\nu)} \tag{3.17}$$

Observe that $C_{1111} \neq E$, unless Poisson's ratio is zero. In the *modulus formulation*, C_{1111} (C_{11} in the reduced notation) is expressed for isotropic solids in terms of the bulk modulus as

$$C_{11} = K + \frac{4}{3}G \tag{3.18}$$

with K as the bulk modulus.

The constant C_{2323} represents a shear modulus since it is the ratio of shear stress to a corresponding shear strain. Specifically, the shear modulus is $C_{44} = G$. Shear moduli may be measured via ultrasonic tests with shear waves, or by quasi-static shear. Similarly, in anisotropic materials C_{55} and C_{66} represent shear moduli that may be different in the corresponding directions. Elastic constants C_{12}, C_{13} link axial stress with strain in an orthogonal direction; they depend upon Poisson's ratio and on axial moduli. Elastic constants such as C_{16} link shear strain with axial stress. This is

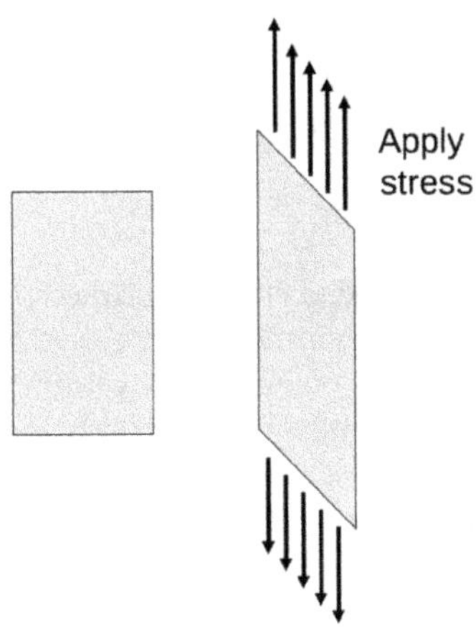

Figure 3.3: Stretch-shear coupling.

called stretch-shear coupling (Figure 3.3). If rotation of the ends is allowed as shown, a meaningful determination of properties can be accomplished. Clamping of the ends would prevent the rotation and cause a complicated pattern of deformation. Elastic constants such as C_{45} link shear stress in one direction with shear strain in another direction. In classical elasticity there is no coupling between stretch and twist, even if there is anisotropy.

For anisotropic solids, a hydrostatic stress can give rise to shear deformation [3]. In the bending of an anisotropic bar [3], the end may have two components of deformation, one orthogonal to the direction expected from the applied moment. A torsional moment upon an anisotropic bar can cause a bending deformation as well as a twisting deformation The stress concentration factor for holes and notches differs from that in an isotropic material. If the anisotropy is large, the stress concentration factor can be much larger or much smaller than for the same shape hole in an isotropic material. The effect depends on the direction of stress with respect to the material principal axes.

3.3.8 Physical interpretation: elastic compliance

The physical interpretation of the constants is more familiar to most people in the *compliance formulation* (Equation 3.4). Compliance is symbolized as J, S, or s. For example, for isotropic materials, $J_{11} = 1/E$ with E as Young's modulus; $J_{12} = -\nu/E$ with ν as Poisson's ratio. The compliance matrix is therefore [14] as follows.

$$\mathbf{J} = \begin{pmatrix} \frac{1}{E} & \frac{-\nu}{E} & \frac{-\nu}{E} & 0 & 0 & 0 \\ \frac{-\nu}{E} & \frac{1}{E} & \frac{-\nu}{E} & 0 & 0 & 0 \\ \frac{-\nu}{E} & \frac{-\nu}{E} & \frac{1}{E} & 0 & 0 & 0 \\ 0 & 0 & 0 & \frac{1}{G} & 0 & 0 \\ 0 & 0 & 0 & 0 & \frac{1}{G} & 0 \\ 0 & 0 & 0 & 0 & 0 & \frac{1}{G} \end{pmatrix} \tag{3.19}$$

Equation 3.19 contains three elastic constants. If they are independent, the material could be cubic in symmetry. For *isotropic* materials, an interrelation must be provided for the constants. Only two elastic constants are independent for isotropic materials; the interrelation $E = 2G(1+\nu)$ applies for isotropic solids. Observe that the matrix in the compliance formulation corresponds with the elementary three-dimensional form of Hooke's law that is familiar.

As for matrix vs. tensor representation in the compliance formulation, factors of two are introduced [6] to account for the factor of two in the distinction between tensorial shear strain and engineering shear: $J_{ijkl} = J_{mn}$ when m and n are 1, 2, 3; and $2J_{ijkl} = J_{mn}$ when either m or n are 4, 5, 6; and $4J_{ijkl} = J_{mn}$ when both m or n are 4, 5, 6. If, following Nye [6], $2(J_{11} - J_{12}) = 1/G$ with G as the shear modulus, there is a factor two difference corresponding to ϵ_4 as elementary shear strain.

As for anisotropic materials, the constant $J_{1111} = 1/E_1$ is the inverse of the Young's modulus E_1 in the 1 direction since all other stress components except σ_{11} are zero. The constant $J_{2211} = -\nu_{12}/E_1$ is related to

Poisson's ratio ν. Poisson's ratios can be different depending on direction. For *orthotropic* materials [14], the compliance matrix is as follows.

$$\mathbf{J} = \begin{pmatrix} \frac{1}{E_1} & \frac{-\nu_{21}}{E_2} & \frac{-\nu_{31}}{E_3} & 0 & 0 & 0 \\ \frac{-\nu_{12}}{E_1} & \frac{1}{E_2} & \frac{-\nu_{32}}{E_3} & 0 & 0 & 0 \\ \frac{-\nu_{13}}{E_1} & \frac{-\nu_{23}}{E_2} & \frac{1}{E_3} & 0 & 0 & 0 \\ 0 & 0 & 0 & \frac{1}{G_{23}} & 0 & 0 \\ 0 & 0 & 0 & 0 & \frac{1}{G_{31}} & 0 \\ 0 & 0 & 0 & 0 & 0 & \frac{1}{G_{12}} \end{pmatrix} \quad (3.20)$$

Poisson's ratio ν_{21} refers to strain in the 2 direction in response to stress in the 1 direction. For orthotropic materials, Poisson's ratio follows a reciprocal relation $\frac{\nu_{ij}}{E_i} = \frac{\nu_{ji}}{E_j}$.

Off-axis elastic constants occur in materials with general anisotropy, including triclinic crystals. They, and monoclinic crystals, exhibit nonzero values of elastic constants such as J_{14} which give rise to shear deformation in response to axial stress. Such coupling can present challenges in the interpretation of experiments. If there is any constraint on the coupled motion, the deformation field and the stress may differ from the assumptions used in the analysis of the experiment.

3.3.9 Physical interpretation: experiment

Experimentally, Young's modulus E is measured in quasi-static tension or compression experiments upon a slender specimen in which stress is applied in one direction and strain is measured in the same direction. Strain in the other directions is free to occur. The measurement of transverse strains reveals the Poisson's ratios. The experiments are most straightforward to interpret in materials of orthotropic symmetry or higher, if the load is applied in a principal direction. Under such conditions, the stretch-shear coupling coefficients vanish. Such experiments reveal the compliance elements J also called S.

Ultrasonic measurements can also reveal the anisotropic elastic constants. Slender specimens are not needed. The wavelength of the longitudinal waves commonly used (at frequency 1 to 20 MHz) is considerably smaller than the typical specimen size, so the Poisson effect is restrained. Determination of longitudinal wave velocity in the principal directions provides the axial C moduli; the velocity of shear waves provides the shear C moduli. Wave velocities in oblique directions reveal the remaining C moduli.

One can also extract the elastic constants from the spectrum of vibration modes in a compact specimen such as a rectangular prism, short cylinder or sphere. The method is called resonant ultrasound spectroscopy (RUS). Computer based inversion is used to obtain the elastic constants from the measured spectrum of mode frequencies.

The anisotropic elasticity formulation can be used for heterogeneous materials provided the (i) length scale of the heterogeneity is sufficiently small and (ii) if there is no systematic variation of properties over the dimensions of the specimen. If the heterogeneity is not small in size, there may arise sensitivity to strain gradients, toughening, and other effects that may be understood in the context of generalized continuum theory as discussed in §8.2. In laminates, there may in fact be an organization of lamina orientation as discussed in §5.3.3. In that case the anisotropic elasticity theory is applied to individual laminae, then lamination theory is used to determine the behavior of the laminate.

3.3.10 How to show the effect of symmetry

The effect of symmetry on the elastic constants is demonstrated as follows. Similar approaches may be used for other physical properties. For example in the direct inspection method [6], if there is a mirror plane so that the $+z$ axis becomes $-z$, the transformation of subscripts is $3 \to -3$ so in the full notation with four subscripts, $11 \to 11$, $22 \to 22$, $33 \to 33$, $13 \to -13$, $12 \to 12$, $23 \to -23$. In the *reduced notation* in which the elastic matrix has two subscripts, $1 \to 1$, $2 \to 2$, $3 \to 3$, $5 \to -5$, $6 \to 6$, $4 \to -4$. So, for example $C_{15} \to -C_{15}$. Because the material has this plane of symmetry, the transformation does not change the components, so $C_{15} = 0$. Similarly $C_{14} = 0$ for this symmetry. By contrast, $C_{44} \to C_{44}$ so this one does not vanish.

The transformation of the modulus tensor may be done directly and explicitly. The transformation of the modulus from an unprimed reference frame to a primed one is

$$C'_{ijkl} = a_{im} a_{jn} a_{ko} a_{lp} C_{mnop} \tag{3.21}$$

with an implied sum on repeated indices. The a are transformation matrices that can represent a rotation or a reflection. For a material with a mirror plane of symmetry so that the z axis becomes $-z$, the transformation matrix is diagonal such that $a_{11} = 1$, $a_{22} = 1$, $a_{33} = -1$, and all other a are zero. It may be written as follows.

$$\mathbf{a} = \begin{pmatrix} 1 & 0 & 0 \\ 0 & 1 & 0 \\ 0 & 0 & -1 \end{pmatrix} \tag{3.22}$$

Consider the stretch-shear coefficient C_{1123}. $C'_{1123} = a_{1m}a_{1n}a_{2o}a_{3p}C_{mnop}$, but the a are diagonal so $C'_{1123} = a_{11}a_{11}a_{22}a_{33}C_{1123} = 1 \times 1 \times 1 \times (-1) = -C_{1123}$. But the material is assumed invariant to reflection so $C'_{1123} = C_{1123}$, so $C_{1123} = 0$; in the reduced notation, $C_{14} = 0$. The process may be repeated for the other elastic constants C_{ijkl}.

3.3.11 Neumann's principle

All symmetry characteristics that show up in physical properties must also show up in the structural point group [6]. The converse does not apply. Some physical properties may not display the full asymmetry present in the underlying structure. For example, chiral asymmetry has no effect on the elastic moduli.

3.4 Stress concentration: anisotropy

The stress on the periphery of a circular hole in a plate under tension stress σ_0 aligned with a principal axis, the x direction, is given as a function of angle θ with respect to the applied stress by [3]

$$\frac{\sigma(\theta)}{\sigma_0} = \frac{E(\theta)}{E_1}[-k cos^2\theta + (1+n)sin^2\theta] \tag{3.23}$$

in which E_1 and E_2 are Young's moduli in principal directions; G is the shear modulus and ν_1 is Poisson's ratio; n depends on elastic constants in a simple way in a principal direction as indicated below. The stress ratio $\frac{\sigma(\theta)}{\sigma_0}$ in Equation 3.23 represents the stress concentration factor, usually taken as the maximum magnitude. The stress concentration factor in classical anisotropic elasticity is independent of hole size.

For arbitrary angle, $E(\theta)$ can be expressed in terms of moduli in a principal direction by rotation of coordinates. If $E_1 = E_{maximum}$, then $k = \sqrt{\frac{E_1}{E_2}}$ and $n = \sqrt{2k + \frac{E_1}{G} - 2\nu_1}$.

For $\theta = 0$, the location is on the diameter aligned with the stress. In the isotropic case, $k = 1$ and $n = 2$ so the stress concentration is -1. For $\theta = 90°$, the location is on the diameter orthogonal to the stress. In the isotropic case, the stress concentration is 3.

For anisotropic materials, the stress concentration factor can substantially exceed the value for the isotropic solid. Even in the cubic case for which $E_1 = E_2$, local stress becomes large if G is sufficiently small.

3.5 Chirality

Chiral materials have an asymmetry in which left and right handedness in the material can be distinguished. Elastic materials which obey Equation 3.2 do not distinguish left from right. An inversion, $x \rightarrow -x$, $y \rightarrow -y$, $z \rightarrow -z$ exchanges right and left forms of a material. A material that is invariant to an inversion is not chiral, also called centrosymmetric.

To show the effect of inversion on the elastic modulus tensor C_{ijkl}, consider the transformation of the moduli under a coordinate change, with $\mathbf{a}$ as the transformation matrix. For an inversion, $a_{im} = -\delta_{im}$ in which the Kronecker delta $\delta_{im} = 1$ if $i = m$; 0 if $i \neq m$. The general transformation is $C'_{ijkl} = a_{im}a_{jn}a_{ko}a_{lp}C_{mnop}$ with an implied sum on repeated indices. The prime denotes the transformed coordinate system. So the only surviving terms in the sum are $C'_{ijkl} = (-1)(-1)(-1)(-1)C_{ijkl} = C_{ijkl}$. The inversion operation therefore has no effect on the elastic modulus tensor.

Anisotropic elastic materials that obey Hooke's law, Equation 3.2, can exhibit stretch-shear coupling (Figure 3.3) but not stretch-twist coupling. Nonetheless, stretch-twist coupling can occur in chiral materials. Figure 3.4 shows stretch-twist coupling in a bar of rubber containing screw-shaped inclusions that provide chirality. To understand such effects, we consider elasticity theory with more freedom than Hooke's law as discussed in §8.4.6.

Chirality is a necessary condition for piezoelectric materials (§4.2.1) to couple mechanical and electrical phenomena. Chirality is also necessary for materials to exhibit optical activity. Optical activity refers to the rotation of the plane of polarization of linearly polarized light as it passes through the material [16] [6] [7]. Directional anisotropy is not necessary for optical activity to occur; only chirality is necessary. A material can be chiral and directionally isotropic, as in liquids or gases containing chiral molecules. Sugar is a chiral molecule; water with dissolved sugar is optically active. Quartz crystals are chiral. Selenium and tellurium form chiral crystals of trigonal symmetry [17].

Chiral chemical compounds, called enantiomers, that constitute mirror-image pairs have been compared with left- hand and right-hand gloves [15]. The chemical and biological behavior of such compounds can depend sensitively on the chirality. The taste, odor, and other aspects of biological activity of compounds is sensitive to their chirality. The word chiral is due to Lord Kelvin in 1893, but studies on optical properties of the chiral crystal quartz were done about 70 years earlier.

Complex chiral structures have been used in making enhanced lenses for electromagnetic waves as discussed in §6.8.2.

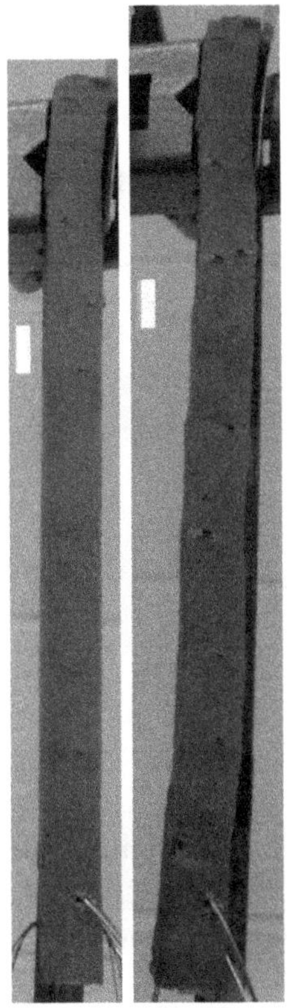

Figure 3.4: Stretch-twist coupling. Left: no tension. Right: tension force 22 N corresponding to a stress 95 kPa. Scale bar: 1 cm.

3.6 Dielectric and optical properties

Electrical properties of materials depend on the symmetry [7]. The dielectric properties may be expressed as the following relation between the electric field vector $\mathcal{E}_k$ and the electric displacement vector $\mathcal{D}_i$,

$$\mathcal{D}_i = K_{ij}\mathcal{E}_j \tag{3.24}$$

in which K_{ij} is the dielectric tensor. For isotropic materials, $K_{ij} = K\delta_{ij}$ with K as a constant and δ_{ij} as the Kronecker delta. The usual dielectric constant k is given by $K = k\epsilon_0$ with ϵ_0 as the permittivity of free space. For anisotropic materials, the dielectric tensor shows direction dependence.

The index of refraction n is given by $n = \sqrt{k}$. If the material is anisotropic, the index of refraction depends on the direction of propagation of the wave and upon its polarization.

3.7 Materials, symmetry and structure

3.7.1 Examples of materials

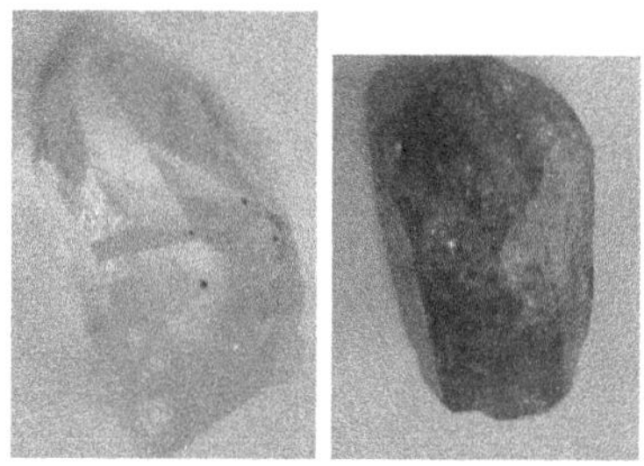

Figure 3.5: Left: quartz crystal has trigonal symmetry. Right: apatite crystal has hexagonal symmetry.

Triclinic crystals have the least symmetry. Examples among minerals include turquoise, $CuAl_6(PO_4)_4(OH)_8\ 4H_2O)$, a semi-precious blue green mineral, and rhodonite, a manganese silicate, (Mn, Fe, Mg, Ca)SiO_3.

Materials with monoclinic symmetry, including single crystals, are invariant to reflections in a single plane of symmetry. Monoclinic crystals include borax (sodium tetraborate) and gypsum (calcium sulfate dihydrate).

Materials with rhombohedral symmetry, also called trigonal, have three equal oblique angles in the unit cell. Quartz in alpha form has trigonal symmetry, Figure 3.5 left; quartz is also chiral; left and right forms exist.

Materials with orthotropic symmetry are invariant to reflections in two orthogonal planes and are describable by nine elastic constants. Materials with axisymmetry, also called transverse isotropy or hexagonal symmetry, are invariant to 60° rotations about an axis and are describable by five independent elastic constants. Apatite, a calcium phosphate, has hexagonal symmetry, Figure 3.5 right. Ice crystals have hexagonal symmetry. As for fibrous materials, a unidirectional fibrous composite may have orthotropic symmetry if the fibers are arranged in a rectangular packing, or hexagonal symmetry if the fibers are packed hexagonally. Wood and bovine plexiform bone have orthotropic symmetry while human compact bone has hexagonal symmetry.

Materials with cubic symmetry are invariant to 90° rotations about each of three orthogonal axes. Cross-ply fibrous composites can exhibit cubic symmetry provided the concentration of fibers in each orthogonal direction is the same. Cubic materials are describable by three independent elastic constants and one thermal expansion coefficient. Cubic single crystals include those of salt, aluminum, iron, and silicon.

Isotropic materials, with properties independent of direction are describable by two independent elastic constants. There are only two independent isotropic elastic constants, hence Young's modulus E, shear modulus G, Poisson's ratio, and also the bulk modulus K are related in an isotropic material. Isotropic materials have only one independent thermal expansion coefficient. Isotropic materials include amorphous solids (such as inorganic glass, glassy polymers, and metallic glass), polycrystalline metals in which the grains are randomly oriented, and composite materials in which the constituents are randomly oriented. Coffee with sugar dissolved in it is isotropic; it is also chiral because the sugar molecules are chiral.

3.7.2 Poisson's ratio in materials with structure

Materials with macroscopic structure visible to the unaided eye include honeycombs and foams (Chapter 6). These materials have been studied using continuum concepts such as stress, strain, elastic modulus and Poisson's ratio though they are clearly not continuous media. No physical material is mathematically continuous: all materials have structure on the atomic scale; almost all materials have structure on larger scales as well. The contrast between the continuum view and the structural view (§1.1.4; Chapter 8) dates back to the foundations of elasticity theory.

A plausible atomic model [18] had been considered to understand elastic behavior. In this model, the atoms are viewed as point particles in a centrosymmetric (not chiral) lattice. The atoms are assumed to interact by central forces dependent upon distance alone. Then Poisson's ratio is 1/4 for all isotropic materials. For that reason the theory was called the uniconstant theory because it predicts isotropic materials have only one independent elastic constant. For anisotropic materials, the tensorial elastic constants are linked by Cauchy relations [18]. Analysis of a 3D isotropic material as a continuum, as is done in the theory of elasticity, allows Poisson's ratio to assume any value between -1 and 0.5. Pioneers such as Navier and Poisson believed the atomic model. The disagreement [19] was ultimately resolved by experiments that showed common isotropic materials had Poisson's ratio close to 1/3 rather than 1/4. The continuum view prevailed. We know that materials do contain atoms but the early model of interatomic interactions was overly restrictive.

In macroscopically homogeneous materials such as polycrystalline metals and amorphous polymers, Poisson's ratio values are linked to the packing of atoms [20]. Poisson's ratio tends to increase with the atom packing density and with atomic number, and to decrease with melting point. Polycrystalline lead, for example, has a Poisson's ratio of about 0.45, and beryllium, 0.03. In materials that can be densified by high pressure, Poisson's ratio tends to increase.

3.7.3 Quasicrystal elasticity

In contrast to conventional crystals, quasicrystals (§3.3.4) are predicted to be elastically isotropic. Quasicrystals of AlCuLi were shown experimentally to be isotropic within 0.07% via resonant ultrasound spectroscopy [21]. Elastic constants were $C_{11} = 112.2$ GPa, $C_{12} = 30.4$ GPa, and $C_{44} = 40.9$ GPa.

3.8 Summary

Symmetry properties of a material determine the kind of anisotropy, if any, that it might exhibit in its physical properties. Anisotropic properties are analyzed using material property tensors which can be displayed in a matrix form. Symmetry classes originally introduced for single crystals are also applicable to other materials such as composites in which the symmetry is manifest on scales larger than that of the arrangement of atoms.

Anisotropy entails direction dependence of properties. Moreover, in anisotropic materials, new effects occur that are not anticipated from study of isotropic materials. These include, for elastic materials, coupling between stretch and shear, and modification of stress concentrations around holes and cracks. Chirality, which involves a distinction between left and right handed microstructure, is essential for properties such as piezoelectricity, presented in the following chapter.

Bibliography

[1] I. S. Sokolnikoff, *Theory of Elasticity*, Krieger, Malabar, FL, (1983).

[2] S. P. Timoshenko and J. N. Goodier, *Theory of Elasticity*, McGraw Hill, New York (1982).

[3] S. G. Lekhnitskii, *Theory of elasticity of an anisotropic body*, Mir, Moscow, (1981).

[4] Y. C. Fung, *Principles of Solid Mechanics*, Prentice Hall, New York (1968).

[5] B. D. Agarwal and L. J. Broutman, *Analysis and performance of fiber composites*, 2nd Ed. J. Wiley, New York, (1990).

[6] J. F. Nye, *Physical Properties of Crystals*, Oxford, Clarendon, (1976).

[7] D. R. Lovett, *Tensor Properties of Crystals*, Adam Hilger, Bristol and Philadelphia, (1989).

[8] D. Shechtman, I. Blech, D. Gratias, and J. W. Cahn, Metallic phase with long-range orientational order and no translational symmetry, Phys. Rev. Lett. 53, 1951-1954, Nov. (1984).

[9] R. Penrose, The role of aesthetics in pure and applied mathematical research, Bulletin of the Institute of Mathematics and its Applications, 10, 266-271 (1974).

[10] R. Penrose, Pentaplexity, Math. Intelligencer, 2, 32-37, (1979).

[11] R. Penrose, US Patent 4133152, Set of tiles for covering a surface, (1979).

[12] `https://en.wikipedia.org/wiki/Crystal_system` .

[13] L. G. Bassett, S. C. Bunce, A. E. Carter, H. M. Clark, H. B. Hollinger,, *Principles of Chemistry*, Prentice Hall, Englewood Cliffs, NJ (1965).

[14] R. M. Jones, *Mechanics of Composites*, Taylor and Francis, Abingdon, UK (1975).

[15] R. Bentley, From optical activity in quartz to chiral drugs: molecular handedness in biology and medicine, Perspect. Biol. Med. 38 (2), 188-229 (1995).

[16] J. C. Bose, On the rotation of plane of polarisation of electric waves by a twisted structure, Proc. Royal Society of London, 63. 146-152, (1898).

[17] D. Royer, and E. Dieulesaint, Elastic and piezoelectric constants of trigonal selenium and tellurium crystals, Journal of Applied Physics 50, 4042-4044 (1979).

[18] J. H. Weiner, *Statistical Mechanics of Elasticity*, J. Wiley, New York (1983).

[19] S. P. Timoshenko, *History of Strength of Materials*, Dover, New York (1983).

[20] G. N. Greaves, A. L. Greer, R. S. Lakes, and T. Rouxel, Poisson's ratio and modern materials, Nature Materials, 10, 823-837 Nov. (2011).

[21] P. S. Spoor, J. D. Maynard, and A. R. Kortan, Elastic isotropy and anisotropy in quasicrystalline and cubic AlCuLi, Phys. Rev. Lett. 75, 3462- 3465, (1995).

Chapter 4

Coupled fields

4.1 Introduction: piezoelectricity, thermoelasticity

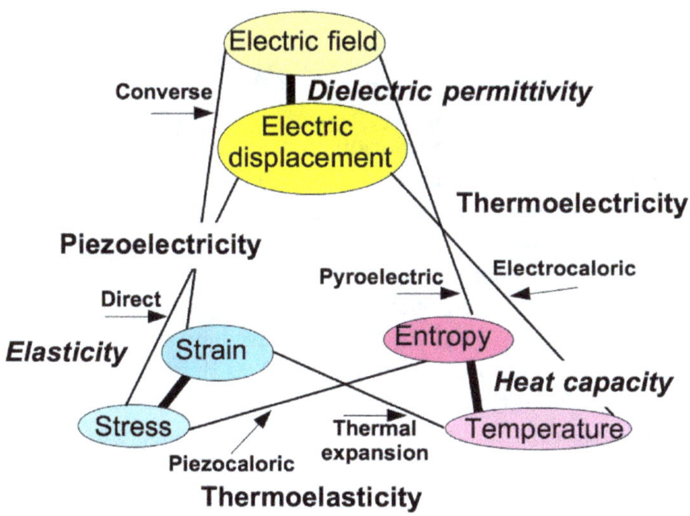

Figure 4.1: Coupled field paradigm for elastic, thermal, and electric variables, adapted from [1], adapted from [2].

Materials have a variety of physical properties, in addition to mechanical properties, that can be described by tensors [1] [3]. For example [1], the strain ϵ_{ij} can depend upon stress σ_{kl} via the elastic compliance J_{ijkl}, upon electric field $\mathcal{E}_k$ (in piezoelectric materials with sensitivity modulus tensor

d_{kij} at constant temperature), and on temperature change ΔT (in most materials, via the thermal expansion α_{ij}). Moreover the electric displacement vector $\mathcal{D}_i$ depends on the electric field via K_{ij} which is the dielectric tensor at constant stress and temperature, and in some materials it can depend on temperature change ΔT via the pyroelectric effect; p_i is the pyroelectric coefficient at constant stress. In Equation 4.3, $\Delta\mathcal{S}$ is entropy change and C is heat capacity per volume. Coupling between elastic, electric and thermal variables is illustrated in Figure 4.1. These phenomena are incorporated in the following linear constitutive equations. Physical properties are taken at constant values of other variables as indicated by the superscripts [1]. For example the compliance $J_{ijkl}^{\mathcal{E},T}$ is the value measured at constant electric field and at constant temperature.

$$\epsilon_{ij} = J_{ijkl}^{\mathcal{E},T}\sigma_{kl} + d_{kij}^{T}\mathcal{E}_k + \alpha_{ij}^{\mathcal{E}}\Delta T \tag{4.1}$$

$$\mathcal{D}_i = d_{ijk}^{T}\sigma_{jk} + K_{ij}^{\sigma,T}\mathcal{E}_j + p_i^{\sigma}\Delta T \tag{4.2}$$

$$\Delta\mathcal{S} = \alpha_{ij}^{\mathcal{E}}\sigma_{ij} + p_i^{\sigma}\mathcal{E}_i + (C^{\sigma,\mathcal{E}}/T)\Delta T. \tag{4.3}$$

Here the usual Einstein summation convention over repeated subscripts is used. Units for the piezoelectric sensitivity d may be expressed as deformation in meters per volt, m/V, or charge per force, coulombs per newton (C/N). The piezoelectric sensitivity and the thermal expansion can be positive or negative. There is no restriction on the sign because there is no energy density associated with cross properties alone. Recall that the elastic moduli were restricted as positive based on an energy density argument. The electric displacement vector $\mathcal{D}_i$ is defined as $\mathcal{D}_i = \epsilon_0\mathcal{E}_i + \mathcal{P}_i$ with $\mathcal{P}_i$ as the polarization and ϵ_0 as the permittivity of free space. The boundary condition on $\mathcal{D}_i$ is pertinent to applications. The change in the normal component of $\mathcal{D}_i$ corresponds to the density of electric charge on the surface.

Piezoelectric properties are presented in §4.2, pyroelectric properties in §4.2.7, thermoelastic properties in §4.3, poro-elasticity in fluid solid composites in §4.4, the Hall effect in §4.5, reciprocity in §4.6, non-reciprocal and extreme materials in §4.6.1, and slow and fast processes including isothermal and adiabatic moduli in §4.7. Thermal expansion of two-phase composites is presented in §5.7 and piezoelectricity of composites is presented in §5.8. Thermal expansion in designed lattices may attain extreme values as shown in §6.7.1.

4.2 Piezoelectric properties

4.2.1 Piezoelectric properties and symmetry

Piezoelectricity differs from elasticity in that to be piezoelectric a material must be *chiral* or non-centrosymmetric: it must lack a center of inversion symmetry. To show this consider the transformation of the moduli under a coordinate change, with **a** as the transformation matrix, a rotation or reflection, $d'_{ijk} = a_{il}a_{jm}a_{kn}d_{lmn}$. The prime denotes the transformed coordinate system. For an inversion, $a_{il} = -\delta_{il}$ (the Kronecker delta, a diagonal matrix) so if the material is invariant to inversions, that is, not chiral, then $d'_{ijk} = (-1)^3 d_{ijk}$. So the inversion changes the sign of all of the d for a non-chiral solid. But it was assumed invariant to the inversions so $d'_{ijk} = +d_{ijk}$. So for a non-chiral solid, $d_{ijk} = 0$. Such a material cannot be piezoelectric.

The cause of piezoelectricity is asymmetry in the charge distribution of atoms or molecules in the material. So chirality is a necessary condition for piezoelectricity but it is not sufficient. There must be a chirality in the charge distribution at some level.

The piezoelectric response d_{ijk} in the full notation is governed by a sensitivity tensor with three indices. The subscripts can take on values from 1 to 3. The tensor elements could be represented in a cubical array but as with the elastic modulus elements it is expedient to use the reduced notation as follows. In the reduced notation, the piezoelectric matrix d_{ij} has two subscripts; the second one takes on values from 1 to 6, so the corresponding matrix is three by six. There can be as many as 18 independent sensitivity coefficients, corresponding to a triclinic crystal.

$$d_{ij} = \begin{pmatrix} d_{11} & d_{12} & d_{13} & d_{14} & d_{15} & d_{16} \\ d_{21} & d_{22} & d_{23} & d_{24} & d_{25} & d_{26} \\ d_{31} & d_{32} & d_{33} & d_{34} & d_{35} & d_{36} \end{pmatrix} \tag{4.4}$$

As an example of a matrix of specific properties, a piezoelectric ceramic composition of lead titanate zirconate, PZT8, has the following properties [4]. The associated symmetry is hexagonal. Because the material is a polycrystalline ceramic, it is better described as transversely isotropic. A manufacturer provides some of the coefficients, with the same values [5].

$$d_{ij} = \begin{pmatrix} 0 & 0 & 0 & 0 & 330 & 0 \\ 0 & 0 & 0 & 330 & 0 & 0 \\ -97 & -97 & 225 & 0 & 0 & 0 \end{pmatrix} pC/N \tag{4.5}$$

This ceramic has relatively large sensitivity coefficients compared with other piezoelectric materials. There is sensitivity to both shear and axial stress components.

Piezoelectricity: cubic symmetry

As with elastic materials, greater symmetry is associated with fewer constants. For example a chiral cubic material must have $d_{11} = d_{33} = 0$, but can exhibit shear sensitivity such as d_{14} in the reduced notation. Recall that this corresponds to d_{123} in the full tensor notation; polarization in the 1 direction couples with a shear stress σ_{23}. The cubic material has three shear entries in the d matrix; these are equal in magnitude: $d_{14} = d_{25} = d_{36}$.

$$d_{ij} = \begin{pmatrix} 0 & 0 & 0 & d_{14} & 0 & 0 \\ 0 & 0 & 0 & 0 & d_{25} & 0 \\ 0 & 0 & 0 & 0 & 0 & d_{36} \end{pmatrix} \tag{4.6}$$

Observe that the only nonzero coefficients correspond to sensitivity to shear stress.

Piezoelectricity: isotropic solids

The permutation symbol e_{ijk} is the only isotropic third rank tensor. Because $e_{ijk} = -e_{ikj}$ isotropic piezoelectric effects entail antisymmetric stress which is forbidden in classical elasticity and in classical piezoelectricity. Asymmetric stress is allowed in generalized elasticity in which the moment from the asymmetry is balanced by a distributed moment as discussed in Chapter 8. In a material of this type, torsional deformation will generate a radial polarization [6].

4.2.2 Piezoelectric materials

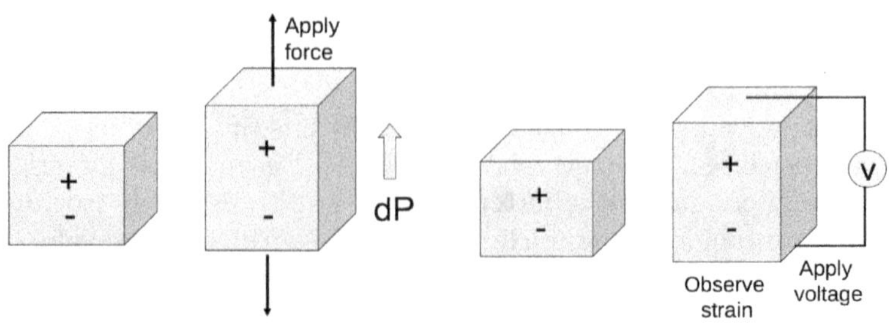

Figure 4.2: Left, piezoelectric direct effect: stress produces polarization by deforming dipoles to generate an increment in polarization dP. Right, piezoelectric converse effect: electric field produces strain by deforming dipoles in the material.

Piezoelectric materials are those which produce an electric polarization when stressed mechanically, as discussed in §4.1. The origin of piezoelectric effects from the deformation of dipoles is shown in Figure 4.2. Piezoelectricity was reported in 1881 by the Curie brothers [7]. The early studies of the Curie brothers were inspired by a prior knowledge of pyroelectric phenomena and an understanding of crystal symmetry [8]. Materials that were initially found to be piezoelectric include quartz, zinc sulfide, tourmaline, topaz, cane sugar and Rochelle salt [8]. Some early authors [8] give piezoelectric constants in CGS ESU units. The conversion is d[esu] = 3×10^4 d[coul/N]. So, for cane sugar, $d_{14} = 0.43$ pC/N and $d_{33} = -3.33$ pC/N.

Rochelle salt or sodium potassium tartrate tetrahydrate ($NaKC_4H_4O_6{\cdot}4H_2O$) has a strong piezoelectric response [9] ($d_{14} = 345$ pC/N at 34°C) which depends on temperature. It is a ferroelectric which has two Curie temperatures, −18°C and 24°C and it disintegrates at 55°C.

Sodium chlorate [10] is a cubic water soluble crystal for which $d_{14} =$ 6.2 pC/N at 28°C. The principal Young's modulus is 45 GPa and Poisson's ratio is 0.23. These properties were determined via a resonance method. The thermal expansion is about 43×10^{-6}/°C. Potassium bromate is also cubic and has piezoelectric properties similar to those of sodium chlorate.

Piezoelectric sensitivity coefficients for various crystals [11] include $d_{15} = 8.2$ pC/N for tourmaline, $d_{33} = 8.6$ pC/N for ZnS, and $d_{15} = 68$ pC/N for lithium niobate $LiNbO_3$ to as much as $d_{14} = 2500$ pC/N for Rochelle salt under different conditions from those above. Because Rochelle salt has a phase transformation in the vicinity of room temperature, properties are highly sensitive to temperature, stress magnitude and other factors including surface layers and quality of mounting of specimens. Experimental imperfections tend to lower the observed value of piezoelectric sensitivity so the highest values reported are likely the most representative of intrinsic properties [8]. Selenium [12] [13] and tellurium [13] trigonal crystals are chiral materials that are piezoelectric semiconductors. Piezoelectric and elastic constants were determined via resonance and wave methods. For selenium, $d_{11} = 41$ pC/N and $d_{14} = 25$ pC/N.

Many of the practical piezoelectric materials are ceramics which exhibit strong effects; piezoelectric polymers are also widely used, and piezoelectric semiconductors are known. Representative properties are given below in Table 4.1. PZT refers to lead titanate zirconate ceramic. Material properties are at ultrasonic frequency. Compliance values such as $s_{33}^{\mathcal{E}}$ are given a superscript $\mathcal{E}$ to denote measurement at constant electric field. This compliance differs from the value measured at constant electric displacement $\mathcal{D}$ due to the coupling between fields. For example $s_{33}^{\mathcal{E}}$ is the inverse of Young's modulus in the 3 direction in the material, measured at constant electric field; $s^{\mathcal{E}}_{44}$ is the inverse of a shear modulus; the 4 direction corresponds to a shear in the 23 direction in the reduced notation used to display properties of anisotropic solids.

Table 4.1: Piezoelectric material properties adapted from [9], at ultrasonic frequency and ambient temperature. Q refers to a mechanical quality factor, the inverse of damping $tan\delta$ (see Chapter 9, §9.2.3); d is a piezoelectric sensitivity, and s is a compliance (also called J). k_3 is the dielectric constant in the 3 direction. PZT refers to lead zirconate titanate ceramic. T_c is the Curie temperature above which piezoelectric properties are lost.

Quantity	Quartz	PZT4	PZT5	$BaTiO_3$	Pb_2NbO_6
d_{33}(pC/N)	0	289	374	190	85
d_{31}(pC/N)	0	−123	−171	−78	−9
d_{11}(pC/N)	2.31	-	-	-	-
d_{15}(pC/N)	-	496	584	260	-
$s^{\mathcal{E}}_{33}$ (TPa^{-1})	9.6	15.5	18.8	9.5	22
$s^{\mathcal{E}}_{11}$ (TPa^{-1})	12.8	12.3	16.4	9.1	-
$s^{\mathcal{E}}_{44}$ (TPa^{-1})	20.0	39	47.5	22.8	-
Q	10^6-10^7	500	75	300	11
T_c (°C)	-	328	365	115	570
k_3	4.68	1300	1700	1700	225

Figure 4.3: Structure of a ferroelectric piezoelectric material, barium titanate, [15], with permission.

Many piezoelectric ceramics are also *ferroelectric*: they have a permanent electric polarization that can be realigned by a sufficiently strong electric field. The permanent electric polarization gives rise to the name *electret*, by analogy to magnets which have permanent magnetization. The

Curie point is a phase transition temperature at which the material changes from a ferroelectric to a non-ferroelectric or paraelectric condition. Ferroelectric materials at temperatures below the Curie point have a complex domain structure (Figure 4.3). Twinning in the crystal structure provides stress relief [14].

The Curie point T_c of these materials refers to the temperature at which they undergo a phase transformation from a ferroelectric material to a less symmetric crystal form, called paraelectric. PZT ceramics are available in a variety of compositions in which different properties are maximized. For example, materials of the highest sensitivity tend to have a comparatively low Curie point, which limits the maximum temperature at which a material can be exposed without depolarization. The Curie temperature is pertinent to high temperature applications. For example, lead metaniobate ($PbNb_2O_6$) has a Curie point of 570°C; it is also available in ceramic form. Single crystal lithium niobate ($LiNbO_3$) has a maximum $d_{15} = 68$ pC/N and a Curie point of 1150°C [11].

The mechanical quality factor Q is the inverse of damping $tan\delta$ as discussed in §9.2.3. Quartz is a crystalline piezoelectric material available in single crystal form. Its viscoelastic loss tangent is extremely small, 10^{-7} at 1 MHz increasing to 10^{-6} at 15 MHz. The low damping is beneficial in resonators used for time keeping and frequency standards. The resonance requires minimal power to maintain, and the resonance curve is has a very sharp peak, so the resonance frequency is well defined. Lead metaniobate ceramic [18], has a high viscoelastic damping $\tan\,\delta = 0.09$ at ultrasonic frequency. The high damping is helpful in reducing reverberation in pulsed ultrasound transducers. Damping is less at audio and sub-audio frequencies [19].

Voltage sensitivity

The above piezoelectric Equations 4.1 and 4.2 are written with stress and electric field as independent variables. One may also express them [9] as strain ϵ_{ij} and electric field $\mathcal{E}_i$ with stress and surface charge density (or electric displacement $\mathcal{D}_j$) as independent variables, assuming constant temperature:

$$\epsilon_{ij} = J^{\mathcal{D},T}_{ijkl}\sigma_{kl} + g^T_{kij}\mathcal{D}_k \tag{4.7}$$

$$\mathcal{E}_i = -g_{ijk}\sigma_{jk} + \beta_{ij}\mathcal{D}_j \tag{4.8}$$

in which β_{ij} is the inverse of the dielectric permittivity K_{ij}.

Large d coefficients are beneficial in applications in which the piezoelectric material is used as an actuator, a driver or an emitter. For sensors in which the output is a voltage, the g coefficients are to be maximized. Because $g = d/K$, materials with a low dielectric permittivity $K = \epsilon_0 k$ (with ϵ_0 as the permittivity of free space) are favored for high g, as shown in Table 4.2. The properties of barium titanate ($BaTiO_3$) ceramic in Table 4.2 differ somewhat from those in Table 4.1 as a result of differences in preparation technique.

Table 4.2: Piezoelectric material properties adapted from [16], at ambient temperature. d and g are piezoelectric sensitivity coefficients, and k is relative dielectric permittivity. $BaTiO_3$ results are for a polycrystalline ceramic. PVDF is a piezoelectric polymer.

Quantity	Quartz	$BaTiO_3$	PVDF
d_{33}(pC/N)	0	191	−7.5
d_{31}(pC/N)	-	−79	0.5
d_{11}(pC/N)	2.3	-	-
$g_{33}(10^{-3}$ VmN$^{-1})$	-	11.4	-
$g_{31}(10^{-3}$ VmN$^{-1})$	-	−4.7	95
$g_{11}(10^{-3}$ VmN$^{-1})$	57	-	-
k_3(-)	4.6	1900	3.1
k_1(-)	4.5	1620	-

Piezoelectric polymers offer thinness, light weight, flexibility, and the ability to be readily formed into intricate shapes. Sheets of the piezoelectric polymer poly(vinylidene fluoride), PVDF are commercially available. This material is ferroelectric, piezoelectric and pyroelectric. Material with sensitivity coefficients of about d_{33} = -20 pC/N, d_{31} = 20 pC/N and g_{33} = 0.15 VmN^{-1} is available commercially [17]. These values are higher than those quoted in Table 4.2 as a result of advances in material preparation. Piezoelectric d coefficients for polymer are about a factor of ten less than for the most sensitive piezoelectric ceramics. Piezoelectric polymers are lead free, are lighter than ceramics, and can be made in flexible films. Piezoelectric polymers offer the further characteristic of a low acoustic impedance, suitable for acoustic applications in which there is an interface with air or water.

4.2.3 Strongly piezoelectric materials

The piezoelectric materials originally available were mineral crystals such as quartz, and materials such as barium titanate. PZT ceramics (Pb(Zr, Ti)O_3), now commonplace, were designed for more sensitive response. Via the Landau theory, a high sensitivity d_{33} can be attained by increasing the dielectric permittivity to flatten the energy function with respect to polarization. Among methods to achieve this goal, operation near a phase transformation boundary is helpful. PZT ceramics were designed to have compositions near the morphotropic phase boundary between the tetragonal and rhombohedral phases. A morphotropic phase boundary is nearly vertical on the phase diagram; properties are much more dependent on composition than on temperature; two phases may coexist over a range of temperature. Compositions near the morphotropic phase boundary exhibit anomalously high dielectric and piezoelectric properties. This is attributed to high polarizability from the coupling between two equivalent energy states. Reducing the Curie temperature further enhances the piezoelectric properties at the expense of a limited temperature range of operation.

Relaxor based ferroelectric single crystals [20] exhibit high piezoelectric coefficients $d_{33} = 1500$ pC/N near phase transition temperatures. Relaxor materials are complex perovskites with a broad spectrum of dielectric dissipation [21]. More recently, relaxor piezoelectric materials [21] such as $Pb(Zn_{1/3}Nb_{2/3})O_3$-$PbTiO_3$ and $Pb(Mg_{1/3}Nb_{2/3})O_3$-$PbTiO_3$ were found to have piezoelectric coefficients (d_{33}) greater than 2500 pC/N. This considerably exceeds that of standard polycrystalline piezoelectric materials such as $Pb(Zr, Ti)O_3$ called PZT. Recently, Sm-doped single crystals of relaxor compositions [22] exhibited even higher d_{33} values ranging from 3400 to 4100 pC/N. High performance is particularly desirable for demanding applications of actuators and sensors. These relaxor single crystals are used in high-frequency medical transducer arrays and other rigorous applications such as underwater sonar [23].

4.2.4 Lead free piezoelectric materials

Lead-free piezoelectric materials are of interest because lead is toxic. The lead in a ceramic is unlikely to harm the user but the manufacture, processing, and disposal of lead containing components can expose people to toxic effects. The well known piezoelectric materials that do not contain lead also do not provide the sensitivity of the PZT ceramics. A lead-free piezoelectric ceramic was reported [24] with d sensitivity similar to that of actuator-grade PZT. The materials were alkaline niobate-based perovskite solid solutions containing niobium, lithium, sodium, potassium and tantalum. Performance of these complex oxides was enhanced by a morphotropic phase boundary.

4.2.5 Experimental piezoelectric measurement

As for experimental methods, piezoelectric d coefficients can be measured by applying a known compressive load, and measuring the resulting polarization using a charge amplifier. The charge amplifier contains a feedback circuit to convert electric charge to a calibrated voltage. This measurement is normally done using a sinusoidal waveform in time. The experimenter may also apply a known voltage waveform and measure the resulting deformation. Because the deformation is on the order of nanometers, the measurement device must be sufficiently sensitive. The sensitivity of piezoelectric materials is at times expressed as a ratio of output voltage to applied stress or in other ways. Piezoelectric properties can also be extracted from study of the structure of the resonant response of a piezoelectric disk to electrical input [9].

4.2.6 Electrostriction

Electrostriction is a quadratic coupling between electric field and strain. It occurs in all materials. It arises from the attraction of surface charges and from the dependence of dielectric constant upon strain. The effect is weak unless high electric fields in excess of 2 kV/mm are applied. Electrostriction has been used in the context of artificial muscles, §4.8.

4.2.7 Pyroelectric materials

In pyroelectric materials, a temperature change gives rise to an electric polarization; temperature, a scalar quantity, is coupled to polarization, a vector, via Equation 4.2. Therefore pyroelectricity requires the material to have chiral asymmetry. There is the further requirement that at least one direction be invariant to all the symmetry operations of the crystal [25]. So the pyroelectric crystal must have either no axis of rotational symmetry or a single axis not included in the inversion symmetry. Consequently piezoelectricity does not entail pyroelectricity. For example, quartz is piezoelectric but not pyroelectric. Pyroelectricity arises in polar crystals in which there is a permanent polarization. This polarization changes with temperature, giving rise to pyroelectricity.

The first known pyroelectric material is the mineral tourmaline. Tourmaline is naturally occurring crystalline boron silicate mineral containing traces of other elements. The earliest observations of heated tourmaline attracting small particles date back more than 2000 years [25]. The term pyroelectricity was introduced and early scientific studies were done more than 190 years ago [26]. Pyroelectric materials include lithium niobate, Rochelle salt, sucrose, tartaric acid, bone and dentin [25]. The piezoelectric ceramics PZT, barium titanate, and lead metaniobate are also pyroelectric; also the piezoelectric polymer poly(vinylidene fluoride), PVF2, also called PVDF. The pyroelectric coefficient of PVDF is $p \approx 30$ μC m^{-2}K^{-1}.

4.2.8 Applications of piezoelectric and pyroelectric solids

Piezoelectric materials are used in transducers including microphones, sound emitters and ultrasonic sensors and emitters, in ink jet printers; also in spark generators for stoves. Piezoelectric materials are used in underwater sonar, in medical imaging devices, and in industrial nondestructive testing. Accelerometers are made by attaching a seismic mass to a piezoelectric bender element. Piezoelectric materials are also used in fine scale positioning devices in optical systems and in scanning probe microscopes such as atomic force microscopes (AFM), and in nano-indenters. Quartz is used in transducers for which low damping (§9.2.3) is desired, in electronic filters, and in frequency standards for radio equipment and timepieces. The high Q and low sensitivity to temperature are advantageous in these applications.

Piezoelectric materials may be used at high temperature as sensors within engines and turbines, in deep drilling, and injection molding. For stable behavior, the Curie point must be well above the operating temperature because anomalies in properties occur as one approaches the Curie point.

Pyroelectric materials are used in thermal sensors and infrared image devices. Electret microphones based on PVDF polymer film make use of permanent polarization to charge a capacitor with an air gap. Sound pressure causes the capacitor spacing to change, generating a time-dependent voltage proportional to sound pressure. Such capacitor or condenser microphones are widely used in computers, cell phones, and other devices. Pyroelectric materials have also been studied to extract energy from thermal fluctuations in the environment: energy harvesting.

4.3 Thermal expansion

4.3.1 Thermoelasticity, symmetry, causes

Thermal expansion of crystalline solids is attributed to the anharmonicity (nonlinearity) of the interatomic potential and is therefore considered to be a property intrinsic to each material. There is no restriction on symmetry; isotropic materials have nonzero thermal expansion. In common materials, expansion is positive and tends to decrease with elastic modulus (Figure 4.4).

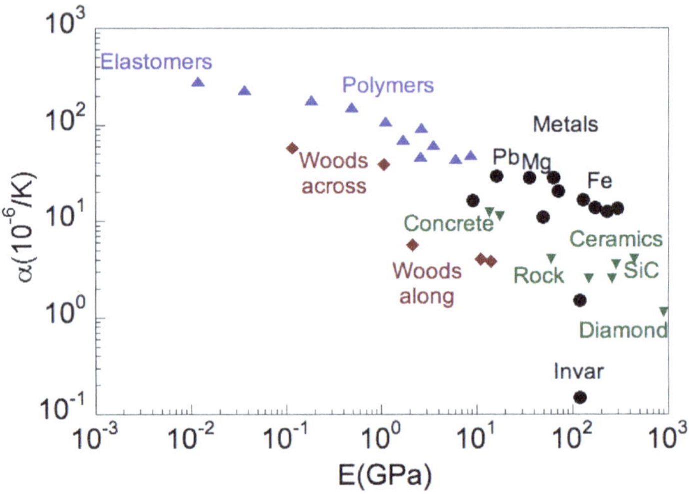

Figure 4.4: Expansion-modulus map of many materials, adapted from [27].

4.3.2 Thermal expansion anisotropy

Thermal expansion is incorporated in the elasticity equations via an additional term containing temperature, Equation 4.1. This may be expressed in matrix form, as is done with the elastic modulus or compliance.

The thermal expansion tensor is symmetric because the strain tensor is symmetric by definition. For a triclinic material, the thermal expansion can contain shear terms as well as axial expansion terms. Because of its symmetry, the expansion tensor α_{ij} may be expressed in a principal axis system. The thermal expansion in a principal reference frame may be written, in the full notation,

$$\alpha_{ij} = \begin{pmatrix} \alpha_{11} & 0 & 0 \\ 0 & \alpha_{22} & 0 \\ 0 & 0 & \alpha_{33} \end{pmatrix} \tag{4.9}$$

So, for the most general anisotropy there are no more than three independent thermal expansion coefficients. In a cubic material, all three principal directions are equivalent. A cubic material therefore has only one thermal expansion coefficient.

Some materials exhibit considerable anisotropy in their thermal expansion. For example, zinc single crystals are hexagonal and have expansions 13 and 64 $\times 10^{-6}/°\mathrm{C}$ in principal directions [1]. In polycrystalline metals with randomly oriented crystallites, the aggregate properties are isotropic.

Even so, a change in the temperature of a material with anisotropic crystals can cause substantial thermal stress between the crystallites. Composite materials may be anisotropic in both elastic and thermal properties. Constituents may have very different thermal expansion values, giving rise to thermal stress between constituents. If, however, the material is cubic or isotropic, the three principal thermal expansions are equal.

Example: thermal expansion under rotation

Consider rotation of a material through an angle θ as expressed by the following matrix.

$$\mathbf{a} = \begin{pmatrix} 1 & 0 & 0 \\ 0 & cos\theta & -sin\theta \\ 0 & sin\theta & cos\theta \end{pmatrix} \tag{4.10}$$

How does the rotation affect the thermal expansion α_{ij}? Show that if the material is orthorhombic, a rotation of coordinates gives rise to nonzero shear terms. Consider specifically α_{32}. Recall that in a principal system the expansion matrix is diagonal.

Solution. The transformation of coordinates is $\alpha'_{ij} = a_{im}a_{jn}\alpha_{mn}$. Recall that repeated indices are summed over.

So $\alpha'_{32} = a_{3m}a_{2n}\alpha_{mn} = a_{33}a_{23}\alpha_{33} + a_{32}a_{22}\alpha_{22}$

$\alpha'_{32} = cos\theta(-sin\theta)\alpha_{33} + sin\theta cos\theta\alpha_{22} = sin\theta cos\theta(\alpha_{22} - \alpha_{33})$.

This shear component is nonzero in general but it vanishes if the angle θ is 0° or 90° or if $\alpha_{33} = \alpha_{22}$.

For comparison, $\alpha'_{33} = a_{3m}a_{3n}\alpha_{mn} = a_{33}a_{33}\alpha_{33} + a_{32}a_{32}\alpha_{22}$, so $\alpha'_{33} = cos^2\theta\alpha_{33} + sin^2\theta\alpha_{22}$. So if $\theta = 0°$, $\alpha'_{33} = \alpha_{33}$; if $\theta = 90°$, $\alpha'_{33} = \alpha_{22}$.

4.3.3 Small or negative thermal expansion

Some materials have a small positive expansion or even a negative expansion. Invar, which is an alloy of Fe and Ni, has a very low expansion [28] that is desirable for materials requiring precise dimensional tolerance or subjected to environments with large fluctuations in temperature. In crystalline materials, negative expansion is often associated with a complex unit cell structure. For example, zirconium tungstate (ZrW_2O_8) is of particular interest because it has a negative expansion [29] over a temperature range [30] of more than 1000°C.

Negative thermal expansion has been observed in other compounds with open framework crystal lattices as reviewed in [32]. Negative thermal expansion effects are interpreted in terms of two components of lattice anharmonicity, one associated with volume change and the other associated with increase in vibration amplitude. Single crystals of $Ag_3Co(CN)_6$ and $Ag_3Fe(CN)_6$ are anisotropic and have a large positive expansion in one direction and a large magnitude of negative expansion in another direction [31]. By contrast, negative expansion in cubic zirconium tungstate is isotropic.

Extreme thermal expansion, positive or negative, of large magnitude can be attained in lattices as presented in §6.7.1; such lattices do not require materials with exotic chemistry.

4.3.4 Composite thermal expansion bounds

As for heterogeneous materials, thermal expansion, as with elastic properties, can be calculated given the constituent expansion values, moduli, and volume fractions. The calculation is straightforward for simple microstructures such as the Voigt and Reuss laminates. Bounds [33] are known for the thermal expansion coefficient of composite materials of two solid phases in terms of constituent expansions. The upper bound is a rule of mixtures or Voigt formula.

$$\alpha_c = \alpha_1 V_1 + \alpha_2(1 - V_1) \tag{4.11}$$

The other bound depends on the bulk modulus K_1, K_2 of each phase and the bulk modulus K of the composite,

$$\alpha_c = \frac{1}{K}[\alpha_1 K_1 V_1 + \alpha_2 K_2(1 - V_1)] \tag{4.12}$$

See, for comparison, expansion of a fibrous material in the longitudinal and transverse directions, §5.7.2.

4.3.5 Applications and thermal expansion

Thermal expansion of materials is of practical interest since materials in service may experience temperatures which vary considerably. Linear expansion becomes macroscopic for large structures. In the case of bridges and roads, it is not practical to use special materials so expansion joints are provided to accommodate the deformation due to thermal expansion. For example, steel used in bridges has an expansion $11 \times 10^{-6}/°\text{C}$. A bridge 2 km long under a temperature change of 60°C would expand 132 mm, clearly enough to require accommodation. If dimensional stability is desired in a smaller scale design, materials of zero or minimal thermal expansion are of interest. Precision instruments in satellites and spacecraft are subject to large temperature excursions; control of thermal expansion is important in such settings. Some alloys have thermal expansion considerably smaller than that of steel or aluminum. Materials that are constrained in a structure or by other constituents in a composite will experience thermal stress when temperature is varied. For such cases as well, it is desirable to minimize expansion itself or the contrast in expansion.

Actuators controlled by temperature changes can make use of structures that exhibit large deformation due to large thermal expansion. Perhaps the most well known example is the thermostat in which deformation of a bi-material strip is used to make or break an electrical contact that controls heaters or coolers, §5.7.3.

4.3.6 Piezocaloric and related effects

The piezocaloric effect, also known as the elastocaloric effect, entails a change in entropy in response to mechanical stress (§4.1). If the process is adiabatic so there is insufficient time for heat to flow, there will be a temperature change. The piezocaloric effect is the converse of thermal expansion. The effect has long been known. Rapidly stretched rubber exhibits a temperature change large enough to easily detect by touching the rubber to one's nose or lips. The earliest report of the effect was in 1805 by Gough [34]. Effects were reported for several materials by Joule [35] in 1859 as adduced in [36].

The electrocaloric effect couples electric field with temperature so that an imposed electric signal causes a change in entropy. If the process is adiabatic so no heat flows, there is a change in temperature. The electrocaloric effect is the converse of the pyroelectric effect in which a temperature change causes the electric polarization to change.

The magnetocaloric effect couples magnetic field with temperature. The effect was used to cool materials to approach absolute zero temperature by the adiabatic demagnetization of a gadolinium compound [37], leading to a Nobel Prize in chemistry. Ordinary refrigerators use compressed gas. Because the gas freezes solid at sufficiently low temperature, other methods of cooling must be used in cryogenic experiments. Magnetocaloric coolers continue to be used in ultra-low temperature laboratories.

Refrigeration via cycles in coupled field effects such as the magnetocaloric effect, electrocaloric effect, and piezocaloric effect remain subjects of current research [36]. Cooling via coupled fields in solids has several potential advantages. Some of the gases used in conventional refrigeration pose environmental concerns. Also, it may be possible to achieve energy efficiency greater than that available in conventional cooling methods.

Coupled field phenomena can greatly increase in magnitude in the vicinity of phase transformations. Studies of such phenomena has been recently reviewed [36].

Stress analysis via thermography

Temperature changes due to the piezocaloric effect are used in an experimental stress analysis method called differential thermography. Small temperature changes from an oscillating stress on an elastic material are observed with a sensitive infrared camera; stress distributions are inferred from the pattern of temperature change. The method [38], called SPATE, which means Stress Pattern Analysis by Thermal Emission, has undergone various refinements [39] [40] [41] and is commercially available. The stress must be time varying or the temperature heterogeneity would disappear via thermal conductivity.

4.4 Fluid-solid composites

4.4.1 Constitutive equations

In a fluid-solid composite, when stress is applied to surfaces, the material deforms, causing fluid to flow through the interstices. Such materials are at times called poroelastic. This behavior can be described by the Biot theory. The Biot [42] theory of stress-induced fluid flow has the same mathematical structure as the theory of thermoelasticity. The Terzhagi [43] theory of fluid flow is a one dimensional special case of the Biot theory, which is three-dimensional. This one dimensional version may be expressed as follows [44], with σ as the stress on the solid phase, and ϵ as the strain. The increment of fluid content ζ is defined as the change in volume ΔV_{fluid} of the fluid divided by the total volume V of the material, so $\zeta = \frac{\Delta V_{fluid}}{V}$. P is fluid pressure. Some authors use a different sign convention.

$$\epsilon = J\sigma + \frac{1}{H}P \tag{4.13}$$

$$\zeta = \frac{1}{H}\sigma + \frac{1}{R}P \tag{4.14}$$

The material coefficients are as follows. J is compliance at constant fluid pressure or drained compressibility, $\frac{1}{H}$ is volume change per pore pressure change, called poroelastic expansion coefficient, and $\frac{1}{R}$, called the specific storage coefficient, relates fluid volume to fluid pressure at constant stress. The symbols correspond to terminology used in geology [44]. In three dimensions, J is the inverse of the bulk modulus and the stress σ is $\sigma_m = \frac{1}{3}(\sigma_{11} + \sigma_{22} + \sigma_{33})$. The strain is interpreted as a volumetric strain.

One needs the shear modulus G of the solid as well to fully describe the material. There is no poro-elastic effect in shear because the shear modulus of fluid is zero.

4.4.2 Experimental determination of constants

Because $J = \frac{\partial \epsilon}{\partial \sigma}|_{P=const}$, the bulk compliance is measured by applying a hydrostatic stress to the solid phase while maintaining the fluid pressure P constant. Because $\frac{1}{H} = \frac{\partial \epsilon}{\partial P}|_{\sigma=const}$, the variable H can be measured by changing the fluid pressure P and measuring the strain in the solid phase while the stress upon the solid is constant. To determine R, a known amount of fluid is extracted under a known pressure on the fluid, with zero stress applied to the solid phase.

To determine the effect of the compressibility J_{fluid} of the fluid, suppose the strain in Equation 4.13 is zero. With $J = 1/K$ and K as the bulk modulus of the material at constant fluid pressure, $\sigma = -\frac{K}{H}P$. Substituting in Equation 4.14, $\zeta = (-\frac{K}{H^2} + \frac{1}{R})P$ so $\frac{\zeta}{P} = (-\frac{K}{H^2} + \frac{1}{R})$. The fluid does not occupy the entire volume, only the pore volume which has volume fraction denoted by the porosity f. So, via the definition of ζ, assuming the pressure does not appreciably change the porosity, $\frac{\zeta}{P} = (-\frac{K}{H^2} + \frac{1}{R}) = fJ_{fluid}$. Recall $\frac{1}{R}$, called the specific storage coefficient, relates fluid volume to fluid pressure at constant stress. So the constant strain condition is seen to differ from the constant stress condition.

Undrained compliance

Suppose the material is deformed and no fluid is allowed to escape. Such a condition may be achieved experimentally by enclosing the material in a chamber with walls much stiffer than the specimen. Then $\zeta = 0$. Then from Equation 4.14, $P = \frac{R}{H}\sigma$. Substitute in Equation 4.13, $\epsilon = (J - \frac{R}{H^2})\sigma$. So $J_{un} = (J - \frac{R}{H^2})$ is the bulk compliance under constraint of no fluid flow. It is called the undrained compliance in contrast to J which is the drained compliance measured under constant fluid pressure P. The difference in bulk compliance may be expressed in a form similar to that for piezoelectric and thermal coupling.

$$J_{un} - J = -\frac{R}{H^2} \tag{4.15}$$

The bulk modulus of water is 2.2 GPa; the bulk modulus of air is 300 kPa for a normal atmospheric pressure of 100 kPa. Water in rocks may contain bubbles of air so one cannot necessarily assume the bulk modulus of pure water.

Waves

Wave speeds in a fluid-solid composite are frequency dependent. At sufficiently high frequency, one can generate in porous media a second type of compressional wave which is slower than the normal compressional wave.

This acoustic slow wave arises from a dynamic interaction between the solid and fluid phases. At ultrasonic frequencies, experimental results have been obtained for fused glass bead media [45] [46] [47] and for bone [48].

4.4.3 Applications: geology and geological engineering

Poro-elasticity is used to understand the motion of water or oil in pores within rock or soil. For example, in the case of water, a volume of 1.7×10^8 m^3 was extracted over a period of years, from a porous rock reservoir, an aquifer [49] $h = 20$ m thick, and $A = 10^9$ m^2 in area. The fluid pressure is reduced by $P = -1$ MPa; the stress on the solid phase is constant. It is observed that there is a vertical contraction of $u = 0.1$ m. This corresponds to the sinking of the land above the porous rock layer. Find the increment of fluid content ζ and the storage coefficient $\frac{1}{R}$.

The strain from the given deformation and initial thickness, is $\epsilon = -0.1$ m / 20 m = - 5 × 10^{-3}. For constant stress, from Equation 4.14, $\zeta = \frac{1}{R}P = \frac{1}{R}(1MPa)$ so $\frac{1}{R} = 5 \times 10^{-9}$ Pa^{-1}. Also, $\zeta = \frac{\Delta V_{fluid}}{V}$. The change in volume of water is given as $\Delta V = 1.7 \times 10^8$ m^3; the volume $V = hA$, the thickness times the area of the reservoir. So, $\zeta = 5 \times 10^{-3}$.

In some regions, extraction of water or oil can cause substantial sinking of the overlying ground surface. For example, extraction of water for agricultural use in the San Joaquin Valley in California [53] caused the ground to sink 9 m by 1981; sinking has continued, in some places by 0.27 m per year [54].

4.4.4 Foams

Foams, §6.4, with communicating pore space are called open-cell foams; if the pores are sealed by cell walls the foam is closed-cell. Flexible polymer foams may have bulk moduli comparable to that of air. The air in the cells of closed-cell foams can contribute substantially to their stiffness. In open-cell foams with small cells the air may take some time to escape when load is applied, giving rise to stiffer behavior for rapid deformation than for slow.

4.4.5 Streaming potentials

Fluid flow in channels in a porous solid can generate electric fields [55]: the electrokinetic effect analyzed by Helmholtz [56] in 1879. The electric signals are called streaming potentials. Conversely if an electric field is applied to a solid that contains fluid filled channels, a pressure difference appears. This pressure difference can be used to pump fluid. After some optimizations it was found that an energy conversion efficiency was at most 0.39% using water as a working fluid in an array of glass tubules. The governing Equations 4.16 and 4.17 contain coupled global variables but have a form similar to equations for coupled fields:

$$\Delta\phi = S\Delta p - i/k_f \tag{4.16}$$

$$v_f = \Delta p/R_f + Si \tag{4.17}$$

in which $\Delta\phi$ is the difference in potential across the material, S is the streaming potential, Δp is the pressure difference, v_f is the fluid flow rate, i is the electric current and k_f is the electrical conductance associated with the fluid. In the absence of an applied potential difference, $\Delta p = -R_f v_f$, so R_f is a viscous resistance to fluid flow. The value of S depends on the viscosity and electrical conductivity and permittivity of the fluid and on the electrochemical properties of the solid phase.

Streaming potentials in geological materials are of interest in earthquake prediction and geothermal prospecting [57]; also in the evaluation of wells for oil and gas exploration [58].

Streaming potential systems were studied [55] [59] in the context of energy harvesting (§4.10).

4.4.6 Vascular materials

Healing and cooling

Recently, micro-vascular materials have been introduced in which the fluid flow can be externally controlled. Research materials have been developed with networks of thin fluid-filled tubes (100 μm or less in diameter). These networks can be used to design self healing materials [60] in which one can pump healing agents such as adhesives or sealants into a crack or other failure. The fluid then polymerizes at the crack, sealing it. In contrast to self healing materials with encapsulated spheres that contain healing agent (§5.2.5), micro-vascular materials allow crack damage in the epoxy polymer to be healed repeatedly. The microvascular network may be created in the polymer substrate via the direct-write assembly of a fugitive organic ink. This is done by depositing the ink layer by layer followed by infiltration with an epoxy resin. The resin is then cured. The ink is removed under a partial vacuum by heating the structure to modest temperatures to liquefy the ink. Microvascular materials, both in their role in healing and in their structure, are considered to be inspired by biological tissues [62]. Microvascular materials (Figure 4.5) have also been considered for applications such as structural damage sensing [61] and cooling of high-performance battery

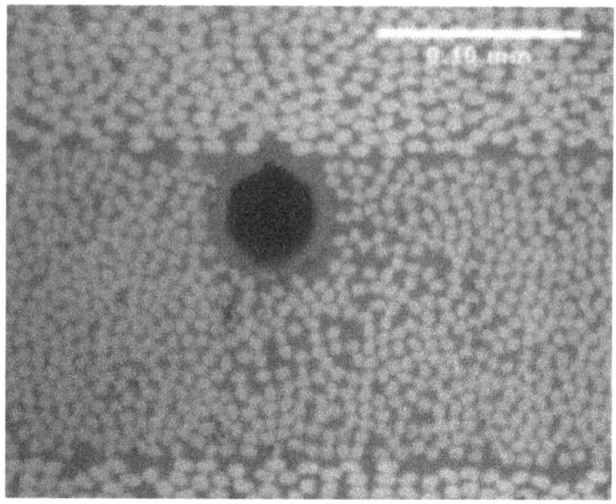

Figure 4.5: Glass-fiber composite with fluid carrying hollow fibers (large dark circle) for self healing and damage sensing [61] with permission. Scale bar: 0.1 mm.

packs [63]. The image is a cross-section; small circles are cross-sections of the glass fibers. The large dark circle is the cross-section of a hollow fiber that can carry fluid. Microvascular materials can also be prepared using sacrificial fibers coated with a release agent; the fibers can be then extracted to leave vascular channels [63].

Laser cooling

A high power laser was designed with a cooling fluid which exhibits a refractive index similar to that of the solid-state laser gain material [64]. The fluid served to cool the laser material and also prevent undesired reflections of light from the interfaces. A mixture of carbon tetrachloride and carbon disulfide was suggested to match the refractive index of neodymium-doped phosphate glass laser elements at the desired infrared wavelength of 1054 nm.

4.5 Hall effect

The Hall effect [65] refers to the action of a magnetic field upon a material carrying an electric current (Figure 4.6). Charge carriers are deflected transversely by the magnetic field to produce a transverse voltage. The Hall coefficient [3] may be written $R_H = \frac{Vt}{BI}$, with B as the magnetic field, I as the current, and t as the thickness of the conductor. In classical physics

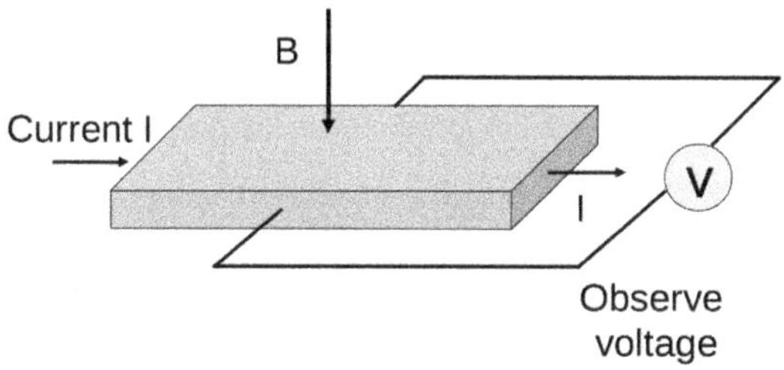

Figure 4.6: Hall effect. Electric current I is deflected by a magnetic field B to generate a voltage V.

[66] [67], the sign of this voltage determines the sign of the charge carrier and its magnitude determines the number n of charge carriers q per volume. R_H in terms of current density j, applied magnetic field B, and resulting electric field $\mathcal{E}_y$ is $R_H = \frac{\mathcal{E}_y}{j_x B_z}$. It may be written $R_H = \frac{1}{nq}$. The Hall effect can be represented by off-diagonal elements of the electrical conductivity tensor that depend upon an applied magnetic field. The Hall effect can occur in isotropic materials. In anisotropic materials the effect depends on direction [3]. In materials of designed complex structure it is possible to control the magnitude and sign of the Hall effect (§6.7.3).

4.6 Reciprocity

Coupling Equations 4.1 and 4.2 are rewritten as follows to show the distinction between the direct piezoelectric effect sensitivity d_{kij}^{dir} (in pC/N), and the converse piezoelectric effect sensitivity d_{kij}^{conv} (in pm/volt): Equation 4.18 and Equation 4.19 respectively. These are different phenomena: polarization in response to stress and strain in response to electric field. Similarly the coefficient of thermal expansion α_{ij}^{exp} and the coefficient of piezocaloric effect α_{ij}^{pzcal} refer to different phenomena.

$$\epsilon_{ij} = J_{ijkl}^{\mathcal{E},T}\sigma_{kl} + d_{kij}^{T:conv}\mathcal{E}_k + \alpha_{ij}^{\mathcal{E}:exp}\Delta T \tag{4.18}$$

$$\mathcal{D}_i = d_{ijk}^{T:dir}\sigma_{jk} + K_{ij}^{\sigma,T}\mathcal{E}_j + p_i^{\sigma}\Delta T \tag{4.19}$$

$$\Delta\mathcal{S} = \alpha_{ij}^{\mathcal{E}:pzcal}\sigma_{ij} + p_i^{\sigma}\mathcal{E}_i + (C^{\sigma,\mathcal{E}}/T)\Delta T \tag{4.20}$$

Reciprocity for piezoelectric solids entails equality between the coefficients for the converse effect and for the direct effect: $d_{kij}^{conv} = d_{kij}^{dir}$. Reciprocity arises from the assumption of equilibrium and the existence of an energy function $dU = \sigma_{ij}d\epsilon_{ij} + \mathcal{E}_i d\mathcal{D}_i + TdS$ [1]. A function of state is defined $\Phi = U - \sigma\epsilon - \mathcal{E}\mathcal{D} - TS$ so $\frac{\partial\Phi}{\partial\sigma} = -\epsilon$ and $\frac{\partial\Phi}{\partial\mathcal{E}} = -\mathcal{D}$.

Further differentiation gives

$$\frac{\partial^2\Phi}{\partial\sigma\partial\mathcal{E}} = -\frac{\partial\epsilon}{\partial\mathcal{E}} \text{ and } \frac{\partial^2\Phi}{\partial\mathcal{E}\partial\sigma} = -\frac{\partial\mathcal{D}}{\partial\sigma}$$

hence the reciprocity relation $\frac{\partial\epsilon}{\partial\mathcal{E}} = \frac{\partial\mathcal{D}}{\partial\sigma}$ for the equality of the direct and converse sensitivity coefficients.

Reciprocity for piezoelectricity is so universally accepted that the same symbol d is routinely used for both sensitivity tensors. There is a similar reciprocity relation between the thermal expansion in Equation 4.1 and the piezocaloric sensitivity in Equation 4.3 in which mechanical stress causes a change in entropy. They refer to different phenomena but they are equal via reciprocity and are given the same symbol α.

Reciprocity in elasticity is formulated in the Maxwell-Betti theorem [68] [69], $F_a u_{ba} = F_b u_{ab}$ in which F_a is the force applied at point a and u_{ba} is the displacement at point a caused by a force at point b; F_b is the force applied at point b and u_{ab} is the displacement at point b caused by a force at point a. The assumptions required to demonstrate elastic reciprocity include time reversal invariance and reversibility.

4.6.1 Non-reciprocal and extreme materials

If energy flux is allowed in the system, extreme stable elastic stiffness is attainable as discussed in §2.5.5. Thermal or electrical energy flux can be modulated to achieve stable high or negative stiffness. In a discrete system of articulated groups of pipes [70] carrying moving fluid, a fluid-solid system (§4.4) achieves stable negative or high values of effective spring constant. Modulated energy flux can facilitate the attainment of a giant piezoelectric response [71] that is non-reciprocal [72]. A flow of heat was used in a composite element containing a material that is piezoelectric and pyroelectric. In such systems the above assumption of existence of an energy function is relaxed.

Nonreciprocal elastic systems [73] were made with an asymmetric hinged rhombus configuration subject to preload that generates nonlinearity.

4.7 Slow and fast processes

4.7.1 Overview

In coupled fields there may be a flow process by which a field variable becomes redistributed with time. Observed properties will depend on whether the measurement is performed slowly or rapidly. Illustrations are given in the following for thermoelastic, piezoelectric, and fluid-solid coupling.

4.7.2 Isothermal and adiabatic moduli

In thermo-elasticity, elastic moduli under adiabatic conditions will differ from moduli under isothermal conditions, as a result of coupling associated with thermal expansion, as follows, with α as thermal expansion, C^{σ} as heat capacity at constant stress and T as temperature.

$$J^S_{ijkl} - J^T_{ijkl} = -\alpha_{ij}\alpha_{kl}\frac{T}{C^{\sigma}} \tag{4.21}$$

Sufficiently slow application of load allows heat generated by thermoelastic coupling to flow out of the material, corresponding to (in the absence of applied gradients) isothermal conditions. If load is applied sufficiently rapidly, there is no time for heat to flow so the experiment is adiabatic. The difference in isothermal and adiabatic moduli can be calculated from the coupling Equations 4.1 and 4.2; for solids it is less than about 2%. The difference in isothermal and adiabatic moduli is analyzed in example §A.1.3.

The rate of heat flow [1] is proportional to the thermal conductivity $k_{ij}^{thermal}$ and the gradient in temperature T.

$$h_i = -k_{ij}^{thermal}\frac{\partial T}{\partial x_j} \tag{4.22}$$

A sudden application of strain to a material will result in a temperature change via the piezocaloric effect. This temperature change will decay with a time constant τ governed by the thermal conductivity, the heat capacity, and the dimensions of the object. The time constant for the bending [74] of a strip of thickness h is $\tau = (\frac{h}{\pi})^2\frac{1}{D}$ with D as the thermal diffusivity. So the time constant is

$$\tau = (\frac{h}{\pi})^2\frac{C_v}{k} \tag{4.23}$$

with k as the thermal conductivity and C_v as the heat capacity per volume.

Sufficiently fast deformation reveals the adiabatic modulus. If the rate of load application is intermediate so it is neither adiabatic nor isothermal, there arises a phase shift between stress and strain, giving rise to mechanical damping (§9.2.3).

As an example, consider a bar of zinc in bending. Determine the thermal time constant for a bar 5 mm thick. The heat capacity per mass $C_v = 383$ J/kg · K is converted to heat capacity per volume (Equation 4.3) via the density $\rho = 7.1\times10^3$kg/m^3. $C_v = (383$ J/kg $\cdot$ K$)(7.1\times10^3$kg/m$^3)$. The thermal conductivity is $k = 113$ J/s $\cdot$ m $\cdot K$.

So, via Equation 4.23, for a bar 5 mm thick, $\tau = 0.061$ sec. If force is applied much more slowly than the time constant, heat is freely transferred across the bar leading to an isothermal condition; the modulus measured is the isothermal modulus. If force is applied much more rapidly, the modulus measured is effectively the adiabatic modulus.

4.7.3 Short-/open-circuit moduli

Elastic moduli of piezoelectric solids depend on the electrical boundary conditions, for example whether the surfaces are in open-circuit or short-circuit conditions. The following difference in compliance can be calculated from the coupling Equations 4.1 and 4.2 and it can be considerable.

$$J^{\mathcal{D}}_{ijkl} - J^{\mathcal{E}}_{ijkl} = -d_{mij}d_{nkl}\frac{1}{K_{mn}} \tag{4.24}$$

The constant $\mathcal{E}$ condition corresponds to a short circuit in which the electric field is forced to zero by electric conduction. The constant $\mathcal{D}$ condition corresponds to an open circuit in which the material is electrically isolated. The boundary condition on $\mathcal{D}$ is that the normal component is continuous across a surface provided there is no free charge. Strictly, the condition $\mathcal{D} = 0$ is obtained if the polarization is normal to the surface.

Consider, for example, the piezoelectric contribution to the compliance of the ceramic PZT 5. From Table 4.1, in the reduced notation, $d_{33} =$ 374 pC/N, $k_3 = 1700$, $K_3 = 1700 \times 8.8\ 10^{-12}$ farad/m. The change in compliance from Equation 4.24 is $J^{\mathcal{D}} - J^{\mathcal{E}} = 9.3 \times 10^{-12}$ m^2/N. But $J^{\mathcal{E}}_{33}$ $= 18.8 \times 10^{-12}$ m^2/N; the inverse of this is the Young's modulus $E^{\mathcal{E}}_3 =$ 53.2 GPa. So $J^{\mathcal{D}}_{33} = 9.5 \times 10^{-12}$ m^2/N; the inverse of this is the Young's modulus $E^{\mathcal{D}}_3 = 105$ GPa, about a factor of two greater than for a short circuit.

Electrical conductivity

The electric current [1] is proportional to the electrical conductivity $k^{electric}_{ij}$ and the electric field $\mathcal{E}_j$.

$$j_i = k^{electric}_{ij}\mathcal{E}_j \tag{4.25}$$

Most piezoelectric materials are excellent insulators that will hold a charge from seconds to minutes. If control of time dependent behavior is desired, an external electrical resistance R may be added. The time constant is then

$$\tau = RC \tag{4.26}$$

with C as the capacitance. Such an adjustable external resistance is used in ultrasonic instruments to control the mechanical damping of the ultrasonic transducer. If electric current is allowed to flow from the piezoelectric object through an electric resistance as force is applied, energy is dissipated. This dissipated energy gives rise to mechanical damping (§9.2.3). The stiffness, hence the natural frequency of vibration of a piezoelectric transducer, may also be tuned by external electrical components.

4.7.4 Fluid-solid composites

The drained compliance J with fluid allowed to escape differs from the undrained compliance J_{un} in which no fluid is allowed to escape, as we have seen in §4.4.2.

$$J_{un} - J = -\frac{R}{H^2} \tag{4.27}$$

Flow q of a fluid of viscosity η in the z direction in a porous medium [44] of permeability k^p is governed by Darcy's law

$$q = \frac{k^p}{\eta}\frac{dP}{dz} \tag{4.28}$$

If the experiment is done sufficiently rapidly, there will be no time for the fluid to escape even if there is no surface constraint.

Permeability depends on the size of the pores. For example, a model material contains tubular channels of diameter d and porosity f. The permeability is $k = \frac{nd^2}{32}$. Rocks may have porosity from 3% to 37% and permeability from more than 3 to less than 10^{-4} darcy [51]. Permeability tends to increase with the square of the pore size in rocks.

Drag due to stress-induced flow gives rise to time or frequency dependent behavior which is manifested as viscoelasticity in the bulk porous material. This is pertinent to geological materials and biological materials.

For Ohio sandstone [50] of porosity 0.191, $G = 6.8$ GPa, $R = 0.33$ GPa, $Q = 0.95$ GPa, $C_{11} = 20.3$ GPa. The permeability of this sandstone is given as 5.6 milli-darcy. Converting to SI units, 1 darcy is equivalent to 0.987×10^{-12} m^2 or 0.987 (μm)^2.

Suppose a load is suddenly applied to a column of material isotropic poroelastic material that is laterally constrained. Fluid is allowed to escape from the top surface through a highly permeable layer. It takes time for fluid to be squeezed from the porous material. The time dependence is described by an infinite series of exponentials in time in which the first term dominates [42]. The time constant τ is

$$\tau = \frac{\eta}{k^p}\left(\frac{\alpha^2}{C_{11}} - \frac{1}{Q}\right)\left(\frac{2h}{\pi}\right)^2 \tag{4.29}$$

with h as the layer height and k^p as the permeability. Here $\alpha = K/H$ and $\frac{1}{Q} = \frac{1}{R} - \frac{K}{H^2}$ with K as the bulk modulus at constant fluid pressure. For isotropic materials the constrained modulus is $C_{11} = K + \frac{4}{3}G$. We remark reference [42] uses an alternate definition of permeability in which the viscosity η is not shown explicitly.

To obtain an order of magnitude of the time constant for a similar kind of stone, assume a laboratory scale specimen of height $h = 0.1$ m, $k = 5$ milli-darcy or 4.94×10^{-15} m^2, and $(\frac{\alpha^2}{C_{11}} - \frac{1}{Q}) = 0.5$ GPa^{-1}. The viscosity is assumed to be that of water at 20°C, $\eta = 1$ milli Pascal second. This value of viscosity may be expressed as 1 centipoise. Then from Equation 4.29, $\tau = 0.41$ sec. If force is applied much more slowly than the time constant, fluid is free to escape; the modulus measured is the drained modulus. If force is applied much more rapidly, the fluid has no time to escape; the modulus measured is effectively the undrained modulus.

4.8 Artificial muscles

Natural muscle can undergo strain from 20% to as much as 40% corresponding to stress of 0.1 MPa to 0.35 MPa [75]. The power density of muscle is 50 to more than 280 W/kg. Muscle can undergo more than 10^9 cycles of operation. Unlike gasoline motors or electric motors, muscle is quiet. The energy density associated with biological metabolism of fat and oxygen is about two orders of magnitude better than that of currently available batteries used for electrical or electromagnetic actuators [76].

The motivation for artificial muscle is that there are no actuators that can replace muscle when it fails or to perform tasks that are easy for living organisms. Artificial muscle could be used for medical implants in replacement for damaged natural muscle and for human assist devices, for minimally invasive surgery, in robotics or in loudspeakers. Piezoelectric materials provide strains that are too small to be used in a muscle-like actuator unless shaped in a rather extreme geometry.

Artificial muscles, presented in 1958, were made using flexible chambers deformed by a pressurized fluid [77]. Artificial muscles based on fluid-solid composites have been used to partially restore function in people with paralysis [78]. The muscle behavior has been modeled [79] [80] using coupled field paradigms.

Electrical actuators made from films of elastomers such as silicone rubber [81] exhibited a maximum strain greater than 100%. The maximum strain and stress was greater than that of natural muscle, and the response time was more rapid. However, the mechanical response is quadratic in the electric field; if linear response is desired, a pre-strain must be applied. The field strength required was from 50 to 400 MV/m; for the films used, voltages were 4 to 6 kV. The force of actuation arises from the attraction of opposite electric charges on the electrodes: electrostriction (§4.2.6).

Shape memory polymers undergo large thermally induced deformations [82] but they usually operate over only one actuation cycle. Polymer gels undergo large deformations in response to changes in the temperature or chemistry of the surrounding fluid. They are, however, weak and slow to respond.

An electrolyte-filled twist-spun carbon nanotube yarn, much thinner than a human hair, functions as a torsional artificial muscle; an electrochemical input gives rise to motion [83]. Artificial muscles were designed based on filled, twist-spun carbon nanotube yarns as electrolyte-free muscles that provide rapid torsional and tensile actuation [84]. Motion could be driven by thermal input or chemical absorption.

Fine scale chirality in wood has been used to make small high-performance actuators that twist in response to a change in relative humidity [85]. Wood based torsional actuators were cut strips of wood 50μm thick and 100-150μm wide; these contained in cross-section several hollow tracheids about 30-50μm wide with helically wound cellulose microfibrils [85]. These actuators were driven by humidity changes.

4.9 Artificial tentacles

Conventional robotic arms mimic the bone and muscle structure of the human arm: rigid links with joints rotated by actuators. Artificial tentacles are an interesting variation.

Soft biological systems such as tentacles are able to move in a complex way. In contrast to muscles, tentacles have multiple contractile elements that allow active bending, twisting and contraction or extension. It is difficult to attempt to replicate these motions with traditional robotic systems. Tentacles can bend and twist anywhere, not just at joints. Soft actuators [86] [87] [88] [89] [90] such as those inspired by octopus tentacles are helpful if they are intended to interact with fragile objects or objects with unknown shape [91]. There is the potential to reduce complexity and cost compared with jointed arm actuators.

A tentacle was made with variable compliance, that can bend independently in two or more regions, and can extend to more than five times its contracted length [86]. It has six degrees of freedom. These artificial tentacles, also called robotic tentacles, or soft robotics, contain embedded multiple flexible chambers in rubber [86] [90]. The chambers are pressurized with air or water to generate motion, for example [93] to achieve deformation similar to that in the heart ventricles. They have been considered for use in medical tests and in surgical tools such as endoscopes. Poro-elastic silicone rubber foams driven by controlled air pressure have been studied for use as soft actuators [92]. The foam was enclosed in a flexible sealing layer. Bending was achieved by attaching a thin strain limiting layer, stiffer than the foam, at one edge. These actuators were used to make a heart-like pump.

4.10 Energy harvesting

Energy harvesting was done in 1981 via a piezoelectric implant in bone [94]. The phrase energy harvesting was not used at that time. The objective was to test for enhanced healing stimulated by stress-induced electrical signals. The study was inspired by the known piezoelectric properties of bone itself (§7.2.2) and the fact that electrical signals stimulate (§7.2.3) healing and growth of bone.

Low power energy harvesting techniques have been evaluated for use in remote applications including in vivo sensors, embedded MEMS (micro electro mechanical systems) devices, and distributed network devices. Materials that exhibit coupled fields are suitable for these applications. Piezoelectric materials offer conversion of mechanical energy to electrical energy but their use in this context [95] is challenging in that their source characteristics including high voltage, low current, high impedance, do not couple well with electronic devices that would use the power. For example piezoelectric films were used to extract electrical power from machine vibration to supply power to micro devices [96]; power was only 5μW. Piezoelectric materials were used to obtain about 1 mW of power at 3 volts from cyclic forces in shoes during walking [97] with the aim of powering wearable electronics. A stack of 145 wafers of PZT was used to generate about 1 mW to power electronics in an experimental knee replacement [95]. Streaming potential systems in which electric signals are generated by fluid flow in a porous medium (§4.4.5) were studied [55] [59] in the context of energy harvesting but efficiency and power levels were low.

Piezoelectric energy harvesters have been studied to power cardiac pacemakers [98]. The rationale was to eliminate the replacement of batteries, a procedure that involves repeat surgery. Piezoelectric energy harvesters have been studied in experimental spinal fusion implants [99]. The rationale was to stimulate healing via electrical signals.

4.11 Other coupled fields

Coupled field variables include magnetic field in magnetic materials, electric resistivity, moisture content, dislocation density, order parameters associated with phase transformations, and the gradient of solute atom concentration in alloys. Coupled fields such as fluid pressure / content in porous materials that contain fluid in the pores entail microscopic or macroscopic heterogeneity. Corresponding coupled field phenomena are known. For example, magnetoresistance refers to a change in electrical resistivity with a change in magnetic field. Materials that have a high value of this coupling are said to exhibit giant magnetoresistance or even colossal magnetoresistance. Such materials are particularly useful in computer disk drives because they convert variations in local magnetic fields in the disk to an electrical signal that enables data to be read.

Humidity in the environment alters the moisture content of wood and some polymers; physical properties can depend strongly on moisture content. Wood (§7.4) becomes more compliant and more time-dependent with increasing moisture. Wood also expands or swells with increasing moisture content.

4.12 Summary

Materials have a variety of physical properties, some of which are coupled. Coupling of elastic and electric fields occurs in piezoelectric materials. Coupling of elastic and thermal fields occurs in all materials with nonzero thermal expansion. Electric and thermal fields are coupled in pyroelectric materials. Piezoelectric and pyroelectric materials require chiral asymmetry in their microstructure. Materials with communicating porosity, if filled with fluid, exhibit poroelastic effects associated with coupling between elastic fields in the solid portion, and pressure or volume change in the fluid portion.

Materials that admit coupled fields form the basis for many applications: sensors, actuators, and active devices. Coupled fields are also pertinent to materials of geological and biological origin.

Bibliography

[1] J. F. Nye, *Physical Properties of Crystals*, Oxford, Clarendon, (1976).

[2] G. Heckmann, The lattice theory of solids, Egeb. exact. Naturwiss. 4, 100-153, (1925).

[3] D. R. Lovett, *Tensor Properties of Crystals*, Adam Hilger; IOP Publishing, Bristol and Philadelphia, (1989).
[4] `http://www.efunda.com/materials/piezo/`
[5] Boston Piezo Optics, `http://bostonpiezooptics.com/`
[6] R. S. Lakes, Third-rank piezoelectricity in isotropic chiral solids, Appl. Phys. Lett., 106, 212905, May (2015).
[7] J. Curie and P. Curie, Contractions et dilatations produites par des tensions dans les cristaux hémièdres à faces inclinées, [Contractions and expansions produced by voltages in hemihedral crystals with inclined faces]. Comptes Rendus (in French). 93, 1137-1140 (1881).
[8] W. G. Cady, *Piezoelectricity*, McGraw Hill, New York (1946); Dover, New York, (1964).
[9] D. A. Berlincourt, D. R. Curran, and H. Jaffe, Piezoelectric and piezomagnetic materials and their function in transducers, in *Physical Acoustics*, ed. E. P. Mason, Vol. 1A, 169-270, (1964).
[10] W. P. Mason, Elastic, piezoelectric and dielectric properties of sodium chlorate and potassium bromate, Phys. Rev., 70, 525-538, (1946).
[11] T. F. Tressler, S. Alkoy, R. E. Newnham, Piezoelectric sensors and sensor materials, J. Electroceramics, 2, 257-272 (1998).
[12] J. Bouat and J. M. Thuillier, Electromechanical resonance in selenium determination of the piezoelectric coefficient d11, Physics Letters A, 37 (1), 71-72, (1971).
[13] D. Royer, and E. Dieulesaint, Elastic and piezoelectric constants of trigonal selenium and tellurium crystals, Journal of Applied Physics 50, 4042-4044 (1979).
[14] G. Arlt, Twinning in ferroelectric and ferroelastic ceramics: stress relief, J. Materials Sci. 25, 2655-2666 (1990).
[15] Liang Dong, Northeast University, China, with permission.
[16] R. E. Newnham, L. J. Bowen, K. A. Klicker, L. E. Cross, Composite piezoelectric transducers, Materials in Engineering, 2 93-106 (1980).
[17] Goodfellow Cambridge Ltd., Ermine Business Park, Huntingdon, England PE29 6WR
[18] G. Goodman, Ferroelectric properties of lead metaniobate, J. Amer. Ceram. Soc. 36, 368-372, (1953).
[19] T. Lee and R. S. Lakes, Damping properties of lead metaniobate, IEEE Transactions on Ultrasonics, Ferroelectrics and Frequency Control, 48, 48-52, (2001).
[20] J. Kuwata, K. Uchino, S. Nomura, Dielectric and piezoelectric properties of 0.91Pb(Zn1/3Nb2/3)O3-0.09PbTiO3 single crystals, Japanese Journal of Applied Physics 21, 1298-1302, (1982).
[21] S. E. Park, T. R. Shrout, Ultrahigh strain and piezoelectric behavior in relaxor based ferroelectric single crystals, J. Applied Physics 82, 1804-1811 (1997).

[22] F. Li, M. J. Cabral, B. Xu, Z. Cheng, E. C. Dickey, J. M. LeBeau, J. Wang, J. Luo, S. Taylor, W. Hackenberger, L. Bellaiche, Z. Xu, L. Chen, T. R. Shrout, S. Zhang, Giant piezoelectricity of Sm-doped Pb(Mg1/3Nb2/3)O3-PbTiO3 single crystals, Science 364, 264-268, (2019).
[23] J. Hlinka, Doubling up piezoelectric performance, Science 364, 228-229, (2019).
[24] Y. Saito, H. Takao, T. Tani, T. Nonoyama, K. Takatori, T. Homma, T. Nagaya and M. Nakamura, Lead-free piezoceramics, Nature 432, 84-87 (2004).
[25] S. B. Lang, *Sourcebook of pyroelectricity*, Gordon and Breach, London, (1974).
[26] D. Brewster, Observations of the pyro-electricity of minerals, The Edinburgh Journal of Science, 1, 208-215 (1824).
[27] M. F. Ashby, On the engineering properties of materials, Acta Metall., 37, 1273-1293, (1989).
[28] C. Woolger, Invar nickel-iron alloy: 100 years on, Mater. World 4(6), 332-333 (1996).
[29] C. Martinek and F. Hummel, Linear thermal expansion of three tungstates, J. Am. Ceram. Soc. 51, 227-228, (1968).
[30] T. A. Mary, J. S. O. Evans, T. Vogt, and A. W. Sleight, Negative thermal expansion from 0.3 to 1050 Kelvin in ZrW2O8, Science 272, 90-92, (1996).
[31] A. L. Goodwin, C. J. Kepert, Negative thermal expansion and low-frequency modes in cyanide-bridged framework materials, Phys Rev B, 71, 140301 (2005).
[32] R. Mittal, M. K. Gupta, S. L. Chaplot, Phonons and anomalous thermal expansion behaviour in crystalline solids, Progress in Materials Science 92 360-445 (2018).
[33] J. L. Cribb, Shrinkage and thermal expansion of a two phase material, Nature 220, 576-577 (1968).
[34] J. Gough, A description of a property of caoutchouc or Indian rubber; with some reflections on the cause of the elasticity of this substance. Mem. Lit. Phil. Soc. Manchester 1 (2nd Series), 288-295 (1805).
[35] J. P. Joule, On some thermodynamic properties of solids. Phil. Trans. 149, 91-131 (1859).
[36] X. Moya, S. Kar-Narayan and N. D. Mathur, Caloric materials near ferroic phase transitions, Nature Materials, 13, 439-450 (2014).
[37] W. F. Giauque and D. P. MacDougall, Attainment of temperatures below 1° absolute by demagnetization of $Gd_2(SO_4)3{\cdot}8H_2O$, Phys. Rev. 43, 768 (1933).
[38] M. H. Belgen, Structural stress measurements with an infrared radiometer, ISA Transactions, 6(1), 49-53 (1967).

[39] D. S. Mountain and J. M. B. Webber, Stress pattern analysis by thermal emission (SPATE). Proc. SPIE (Fourth Euro Electro-Optics Conf.), 164, 189-196 (1978).
[40] J. R. Lesniak and B. R. Boyce, A high speed differential thermographic camera. Proc. SEM Spring Conference on Experimental Mechanics, Baltimore, MD, 491-497 (1994).
[41] P. Stanley, Beginnings and early development of thermoelastic stress analysis, Strain 44, 285-297 (2008).
[42] M. A. Biot, General theory of three-dimensional consolidation, J. Applied Physics 12, 155-164, (1941).
[43] K. Terzhagi, *Theoretical Soil Mechanics*, J. Wiley, New York, (1943).
[44] H. Wang, *Theory of linear poroelasticity*, Princeton University Press, Princeton NJ, (2000).
[45] T. J. Plona, Observation of a second bulk compressional wave in a porous medium at ultrasonic frequencies, Applied Physics Letters 36, 259-261, (1980).
[46] J. G. Berryman, Confirmation of Biot's theory, Applied Physics Letters 37, 382-384, (1980).
[47] D. L. Johnson, T. J. Plona, Acoustic slow waves and the consolidation transition, J. Acoust. Soc. Am., 72, 556-565, (1982).
[48] R. S. Lakes, H. S. Yoon, and J. L. Katz, Slow compressional wave propagation in wet human and bovine cortical bone, Science, 220 p. 513-515, (1983).
[49] E. Meinzer, Compressibility and elasticity of artesian aquifers, Economic Geology, 23(3), 263-291 (1928).
[50] C. H. Yew and P. N. Jogi, The determination of Biot's parameters for sandstones, Experimental Mechanics 167-172 (1978).
[51] J. C. Jaeger and N. G. W. Cook, *Fundamentals of rock mechanics*, Chapman and Hall, London, 3rd edition (1979).
[52] J. Bear, *Dynamics of fluids in porous media*, Dover, New York (1972).
[53] R. L. Ireland, Land subsidence in the San Joaquin Valley, California, as of 1983, US Geological Survey Water-Resources Investigations Report 85-4196, (1986).
[54] M. Sneed and J. T. Brandt: Land subsidence in the San Joaquin Valley, California, USA, 2007-2014, Proc. IAHS (International Association of Hydrological Sciences), 372, 23-27, (2015).
[55] J. F. Osterle, Electrokinetic energy conversion, J. Appl. Mech. 31, 161-164, (1964).
[56] H. L. F. Helmholtz, Studien uber electrische Grenzschichten, Annalen der Physik und Chemie, 7, 337-382 (1879).
[57] F. D. Morgan, E. R. Williams, T. R. Madden, Streaming potential properties of westerly granite with applications, J. Geophysical Research, 94, 12449-12461 (1989).

[58] US patent 1913293, C. Schlumberger, Electrical process for the geological investigation of the porous strata traversed by drill holes (1933).
[59] W. Olthuis, B. Schippers, J. Eijkel, A. van den Berg, Energy from streaming current and potential, Sensors and Actuators B. 111-112, 385-389 (2005).
[60] K. S. Toohey, N. R. Sottos, Lewis, J. A., Moore, J. S., and S. R. White, Self-healing materials with microvascular networks. Nature materials, 6(8), 581-585 (2007).
[61] S. C. Olugebefola, A. M. Aragon, C. J. Hansen, A. R. Hamilton, B. D. Kozola, W. Wu, and S. R. White, Polymer microvascular network composites, Journal of composite materials, 44(22), 2587-2603 (2010).
[62] S. C. Olugebefola, A. R. Hamilton, N. R. Sottos, and S. R. White, Self-healing of internal damage in synthetic vascular materials, Adv. Mater., 22, 5159-5163 (2010).
[63] S. J. Pety, P. X. L. Chia, S. M. Carrington and S. R. White, Active cooling of microvascular composites for battery packaging, Smart Mater. Struct. 26 105004 (11pp) (2017).
[64] M. D. Perry, P. S. Banks, J. Zwiback, R. W. Schleicher, U.S. Patent 7,366,211 B542, 2, Laser containing a distributed gain medium, (2008).
[65] E. Hall, On a new action of the magnet on electric currents, American Journal of Mathematics, 2(3), 287-292 (1879).
[66] D. Halliday and R. Resnick, *Physics*, vol. II, J. Wiley, New York, (1962).
[67] D. W. Preston and E. R. Dietz, *The Art of Experimental Physics*, J. Wiley, New York, (1991).
[68] J. C. Maxwell, On the calculation of the equilibrium and stiffness of frames. Philos. Mag. Series 4, 27, 294-299 (1864).
[69] E. Betti, Teoria della elasticita, Nuovo Cimento 7, 69-97 (1872).
[70] J. M. T. Thompson, 'Paradoxical' mechanics under fluid flow, Nature, 296, 135-137 (1982).
[71] R. S. Lakes, Giant enhancement in effective piezoelectric sensitivity by pyroelectric coupling, EPL (Europhysics Letters), 98, 47001 May (2012).
[72] D. Faust and R. S. Lakes, Reciprocity failure in piezoelectric polymer composite, Physica Scripta, 90 085807 (2015).
[73] C. Coulais, D. Sounas and A. Alu, Static non-reciprocity in mechanical metamaterials, Nature, 542, 461-465 (2017).
[74] C. Zener, Internal friction in solids I - Theory of internal friction in reeds, Phys. Rev. 52, 230-235, (1937).
[75] R. Full and K. Meijer, Metrics of natural muscle, in *Electro Active Polymers (EAP) as artificial muscles, reality potential and challenges*, ed. Y. Bar-Cohen, SPIE Press, Bellingham, Washington (2001).

[76] J. D. W. Madden, N.A. Vandesteeg, P. A. Anquetil, P. G. A. Madden, A. Takshi, R. Z. Pytel, S. R. Lafontaine, Artificial muscle technology: physical principles and naval prospects, IEEE Journal of Oceanic Engineering 29(3), 706-728, (2004).
[77] R. H. Gaylord, Fluid actuated motor system and stroking device, United States Patent 2,844,126, (1958).
[78] V. L. Nickel, J. Perry, and A. L.Garrett, Development of useful function in the severely paralyzed hand, Journal of Bone and Joint Surgery, 45A, (5), 933-952 (1963).
[79] H. F. Schulte, The characteristics of the McKibben artificial muscle, In: *The Application of External Power in Prosthetics and Orthotics*, Publication 874, National Academy of Sciences - National Research Council, Washington DC, Appendix H, 94-115 (1961).
[80] G. K. Klute, B. Hannaford, Accounting for elastic energy storage in McKibben artificial muscle actuators, ASME Journal of Dynamic Systems, Measurement, and Control, 122(2), 386-388, (2000).
[81] R. Pelrine, R. Kornbluh, Q. Pei, J. Joseph, High speed electrically actuated elastomers with strain greater than 100%, Science 287, 836-839 (2000).
[82] A. Lendlein and R. Langer, Biodegradable, elastic shape-memory polymers for potential biomedical applications, Science, 296, 1673-1676 (2002).
[83] J. Foroughi, G. M. Spinks, G. G. Wallace, J. Oh, M. E. Kozlov, S. Fang, T. Mirfakhrai, J. D. W. Madden, M. K. Shin, S. J. Kim, R. H. Baughman, Torsional carbon nanotube artificial muscles, Science 334 494-497 (2011).
[84] M. D. Lima, N. Li, M. de Andrade, S. Fang, J. Oh, G M. Spinks, M. E. Kozlov, C. S. Haines, D. Suh, J. Foroughi, S. J. Kim, Y. Chen, T. Ware, M. K. Shin, L. D. Machado, A. F. Fonseca, J .Madden, W. Voit, D. S. Galvao, R. H. Baughman, Electrically, chemically, and photonically powered torsional and tensile actuation of hybrid carbon nanotube yarn muscles, Science, 338, 928-932 (2012).
[85] N. Plaza, S. L Zelinka, D. S. Stone and J. E Jakes, Plant-based torsional actuator with memory, Smart Materials and Structures, 22(7), 072001 (2013).
[86] G. Immega, K. Antonelli, The KSI tentacle manipulator, Proceedings of 1995 IEEE International Conference on Robotics and Automation (1995).
[87] H. Takagi and Y. Nishi, Flexible Robot Arm, U.S. Patent 5,174,168, (1992).
[88] G. Immega, Bellows Actuator, U.S. Patent 5,181,452, (1993).
[89] G. Immega, Tentacle-Like Manipulators with Adjustable Tension Lines, U.S. Patent 5,317,952, (1994).

[90] M. B. Pritts, C. D. Rahn, Design of an artificial muscle continuum robot. in Proc. IEEE Int. Conf. Robotics Automation (ICRA 04), IEEE, p. 4742 (2004).
[91] C. Laschi, M. Cianchetti, B. Mazzolai, L.Margheri, M. Follador, P. Dario, Soft robot arm inspired by the octopus, Adv. Robot. 26 709-727 (2012).
[92] B. C. MacMurray, X. An, S. S. Robinson, I. M. van Meerbeek, K. W. O'Brien, H. Zhao, R. F. Shepherd, Poroelastic foams for simple fabrication of complex soft robots, Adv. Mater. 27 6334-6340 (2015).
[93] E. T. Roche, R. Wohlfarth, J. T. B. Overvelde, N. V. Vasilyev, F. A. Pigula, D. J. Mooney, K. Bertoldi, and C. J. Walsh, A bioinspired soft actuated material, Adv. Mater. 26 1200-1206 (2014).
[94] J. B. Park, B. J. Kelly, G. H. Kenner, A.F. von Recum, M. F. Grether, W. W. Coffeen, Piezoelectric ceramic implants: In vivo results, J. Biomed. Mater. Res., 15 , pp. 103-110, (1981).
[95] S. R. Platt, S. Farritor, and H. Haider, On low frequency electric power generation with PZT ceramics, IEEE/ASME Transactions on Mechatronics 10(2), 240- 252, (2005).
[96] P. Glynne-Jones, S. P. Beeby, and N. M. White, Towards a piezoelectric vibration-powered microgenerator, IEE Proc. Sci. Meas. Technol., 148(2) 68-72, (2001).
[97] N. S. Shenck and J. A. Paradiso, Energy scavenging with shoe-mounted piezoelectrics, IEEE Micro, 21(3), 30-42, (2001).
[98] M. H. Ansari and M A. Karami, Modeling and experimental verification of a fan-folded vibration energy harvester for leadless pacemakers, Journal of Applied Physics, 119, 094506 (2016).
[99] N. C. Goetzinger, E. J. Tobaben, J. P. Domann, P. M. Arnold, and E. A. Friis, Composite piezoelectric spinal fusion implant: effects of stacked generators, J Biomed. Mater. Res., Part B Applied Biomaterials. 104(1), 158-164 (2016).

Chapter 5

Particles, fibers, platelets

5.1 Introduction: structure

Composites used in engineering applications may have a particulate, fibrous [1] [2], platelet, or cellular structure. In contrast to the materials analyzed in §2.4, the inclusions are not always dilute or random in orientation. Moreover, while theory indicates the possibility of attaining the bounds on moduli, it is not always practical to do so. Particles may be rounded (Figure 2.1), nearly spherical or jagged in shape (Figure 5.1). Practical particulate composites (§5.2) commonly have a distribution of particle sizes, either by design or unintentionally (Figure 5.1). Fibers are long in one direction. Practical fibrous composites (§5.3) commonly have a low order of hierarchical structure [3] in which fibers are embedded in a matrix to form an anisotropic sheet or lamina; such laminae are bonded to form a laminate (Figure 5.3). In the analysis of fibrous composites, the fibers and matrix are regarded as continuous media in the analysis of the lamina; the laminae are then regarded as continuous in the analysis of the laminate. The stacking sequence of laminae and the orientation of fibers within them governs the composite anisotropy. Platelets (§5.4) are long in two directions and thin in a third direction. Platelets may be organized randomly or aligned in layer form.

Waves in composites are discussed in §9.3.5 because attenuation of waves is pertinent. Control of waves in materials with void space is discussed in §6.8. The effect of negative moduli on waves in composites is discussed in §2.5.4.

Composites are useful in enhancing properties other than mechanical. For example in piezoelectric composites (§5.8) some sensitivity properties may be greater than that of either constituent. Thermal expansion (§5.7) of practical composites may exhibit considerable anisotropy.

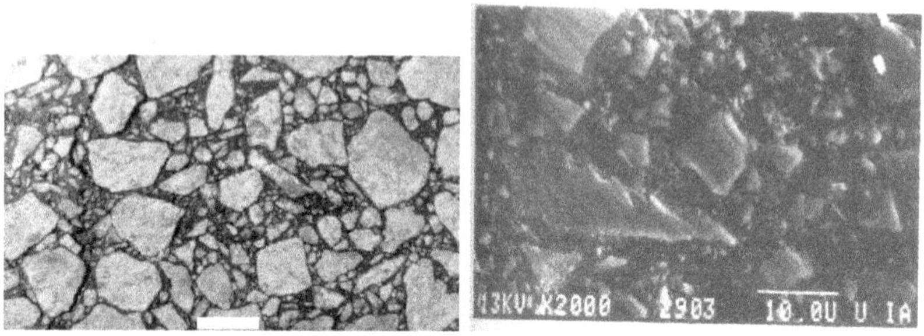

Figure 5.1: Structure of particulate composites; note difference in scale. Asphalt (left); scale bar 10 mm. Dental composite (right); scale bar 10μm, [4], with permission.

5.2 Particulate polymer matrix solids

In many particulate composites [6] [7] [8], a polymer matrix is reinforced by stiff inclusions. The stiffness of a composite with spherical inclusions, as a function of volume fraction of inclusions, is experimentally found to be close to the Hashin-Shtrikman lower bound for isotropic materials, provided the concentration is not too high.

5.2.1 Dental composites

Many different materials have been used as alternatives to amalgam or gold for restorations of cavities in teeth. Resin composites, when they were first developed, were introduced into dental practice as esthetic restorative materials for anterior teeth. However, the growing demand for more esthetic restorations and minimal loss of tooth substance in cavity preparations has made posterior composites an attractive alternative to amalgams. Packable composites are characterized by a high filler load and a filler distribution that gives them a stiff consistency. They are recommended for stress bearing posterior restorations because they have improved handling properties, and can be applied using a technique similar to that used for amalgam.

Dental composite filling resins (Figure 5.1) contain particles of stiff mineral such as barium glass or silica, in a polymer matrix [10] [11]. The silica inclusions confer stiffness and abrasion resistance superior to the polymer matrix alone and approaching the durability of the tooth. The matrix may consist of polymers such as BIS-GMA, an addition reaction product of bis (4 -hydroxyphenol), dimethylmethane, and glycidyl methacrylate. Since the material is mixed, then placed in the prepared cavity to polymerize, the viscosity must be sufficiently low and the polymerization must be controllable. Low viscosity liquids such as triethylene glycol dimethacrylate

(TEGDMA) are used to lower the viscosity and inhibitors such as BHT (butylated trioxytoluene, or 2,4,6-tri-tert-butylphenol) are used to prevent premature polymerization. Polymerization can be initiated by a thermochemical initiator such as benzoyl peroxide, or by a photochemical initiator (benzoin alkyl ether) which generates free radicals when subjected to ultraviolet light from a lamp used by the dentist.

Dental composites typically have a Young's modulus of 10–16 GPa, and a compressive strength of 170–260 MPa. In view of the greater density of the inorganic filler phase, a 77% weight percent of filler corresponds to a volume percent of about 55%. Even though these materials contain a high concentration of filler, significant long term creep, corresponding to a change of a factor of four in stiffness with time, can occur [12]. As with other dental restoration materials, the thermal expansion of these materials exceeds that of tooth structure. The difference is thought to contribute to the leakage of saliva, bacteria, etc, at the interface margins. Use of colloidal silica in microfilled composites allows these resins to be polished, so that less wear occurs and less plaque accumulates. It is more difficult, however, to make these with a high fraction of filler, since the tendency for high viscosity of the un-polymerized paste must be counteracted. An excessively high viscosity is problematical since it prevents the dentist from adequately packing the paste into the prepared cavity; the material will then fill in crevices less effectively.

Dental composite resins have become established as restorative materials for both anterior and posterior teeth [13]. The use of these materials is likely to increase as improved compositions are developed and in response to concern over long term toxicity of silver-mercury amalgam fillings.

Particles on the nanoscale are used in dental composites to achieve improved properties [14]. Hybrid composites, microhybrids and nanohybrids contain a broad distribution of particle sizes. A wide distribution of particle sizes can lead to high filler concentration with resultant high strength and wear resistance. Nano-cluster structure reduces the spacing of the filler particles, allowing the designer to achieve higher concentrations of particles. Fillers are used that are a combination of non-agglomerated and non-aggregated 20 nm silica filler, non-agglomerated and non-aggregated 4 to 11 nm zirconia filler, and aggregated zirconia / silica cluster filler (comprised of 20 nm silica and 4 to 11 nm zirconia particles. The strength for this composite is 160 MPa in flexure, 370 MPa in compression, and 90 MPa in tension (via the diametral method). Young's modulus is from 9 to 11 GPa depending on optical properties. Optical properties are pertinent to dental composites that are intended to mimic the multiple functions of the tooth. Tooth stiffness, strength and wear resistance relate to the function in chewing; optical properties relate to aesthetic qualities. Designers consider color, opacity, translucence and opalescence as well as mechanical properties [15].

As for root canals, endodontically treated teeth are weakened because of dental caries, removal of older restorations and because of the endodontic procedure. Placement of a root canal post causes further loss of tooth substance, therefore making their use favorable only in cases with adequate tooth volume. Until recently, metal posts were the only material available in the restoration of endodontically treated teeth and they exhibited satisfactory clinical results. However, the increased contemporary aesthetic demands led to the introduction of composite posts as a more aesthetic alternative to metal posts. These posts are fiber-reinforced composites (FRC) and consist of unidirectional fibers impregnated in a polymer matrix. The polymer matrix is usually made of a highly cross-linked epoxy resin with a high degree of conversion which protects and supports the fibers while distributing the stresses inside the composite. The polymer matrix retains the orientation of the fibers and determines the shear strength and temperature limitations of the composite material. The fibers can be made of carbon, glass or quartz and they increase the elastic modulus.

5.2.2 Asphalt

Asphalt as used in roads is a particulate, porous composite (Figure 5.1, left). The particles are gravel (2–20 mm in size), sand (0.08 mm – 2 mm), and finely ground stone (1–80 μm) [9]. The stone used in the particles must be sufficiently strong. Too, different types of stone exhibit different adhesion to the matrix; limestone adheres better than silica-based stone [16]. The volume fraction of inclusions may vary widely, up to 85%, corresponding to about 95% by weight. There may be as much as 7% porosity from air-filled pores unintentionally introduced during mixing. Porosity is reduced during compaction and usage. The matrix, known as binder, is a petroleum derivative. Its density is from 0.9 to 1.02 g/cc [16]. There is a wide range of molecular weights. Asphalt binder behaves as a viscoelastic liquid or un- cross linked polymer with a shear modulus of about 1 GPa in the glassy regime of high frequency and low temperature. At higher temperatures and longer loading times, the asphalt becomes softer and behaves more like a viscous fluid. At lower temperatures and for rapid loads, the asphalt becomes stiffer and more elastic. Asphalt also undergoes age hardening in which stiffness increases with the time following preparation.

The viscoelastic behavior is dominated by that of the matrix since asphalt is a composite with stiff particles. The constituent moduli differ a great deal, therefore the bounds on properties are far apart. Since the particles are irregular, even angular, finite element analysis is helpful [17].

Moreover, the particles have a large concentration, so dilute approximations are inadequate for the full range of concentration / volume fraction. As for strength, strength increases with the volume fraction of particles, up to about 80%.

5.2.3 Toughened polymers

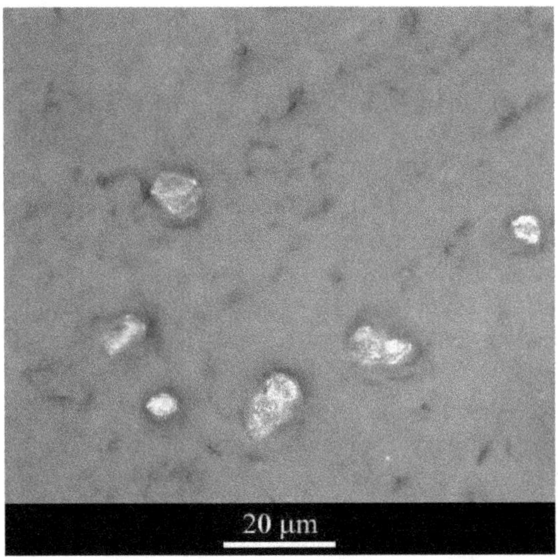

Figure 5.2: Structure of high impact polystyrene; scale bar 20μm, [5], with permission.

The toughness of stiff polymers is enhanced substantially, by more than a factor of ten, by incorporating 5% to 35% particles of rubber by weight [18]. The constituent materials may be combined by physical blending at a temperature above the softening point of the brittle matrix. The rubber may also be dissolved in the monomer from which the matrix is made. As the matrix is polymerized, the rubber precipitates as spherical inclusions. The toughening effect is attributed to crazing and shear yielding that is brought on by the local stress field near the rubber inclusions. A craze is a small crack-like defect that forms in large numbers; it differs from a crack in that it continues to support substantial load. Crazing in transparent polymers is visible as a whitening effect. The modulus of the rubber-modified polymer is reduced by the presence of the soft rubber. The reduction in modulus is modest. Spherical inclusions are less effective than other shapes such as fibers or flakes in altering the modulus, either upward in the case of stiff inclusions, or downward in the case of soft inclusions.

Among toughened polymers, high impact polystyrene is perhaps the best known. High impact polystyrene (HIPS) or rubber-toughened polystyrene is a two-phase mixture of polybutadiene (a synthetic rubber) in polystyrene. The two phases are immiscible. Polybutadiene crystallizes out of the melt to form spheroidal particles as shown in Figure 5.2.

5.2.4 Filled polymers; tire rubber; nano-fillers

Rubber is often reinforced with particulate filler. Tires, for example, contain at least 50% by weight of carbon black particles. These particles improve the strength by as much as a factor of ten, and also improve the abrasion resistance. Carbon black is produced by incomplete combustion of heavy petroleum distillates. If a color other than black is desired, such as in white tennis shoes, silica is used as a reinforcing inclusion. Silica-based fillers are also used in tires because they can provide a lower energy loss in rolling compared to carbon black-filled tires.

Fillers consisting of comparatively large particulate inclusions in a more compliant polymer matrix tend to increase the modulus and have little effect on the damping in agreement with composite theory. If the inclusions are sufficiently small, phenomena occur on the molecular scale, not anticipated in composite theory. For example, rubber filled with carbon black inclusions of diameter on the order 30 nm exhibits viscoelastic properties quite different from those of unfilled rubber [19]. The filler has a minimal effect in the glassy regime but it shifts the α transition to a lower frequency by about one decade (a factor of ten in frequency). The equilibrium modulus (at long time or high temperature) is markedly increased by the filler. This is understood by comparing the inclusion separation 10 nm in a heavily filled rubber, with the similar distance between cross-links in the rubber in question. The filler particle acts as a further cross-link between molecules. Filled rubber tends to be highly nonlinear, in which high amplitude dynamic strain reduces the stiffening effect of the filler [20].

Fine grain inclusions may behave differently from larger scale inclusions due to scale-dependent interactions and molecular scale phenomena. Such inclusions have been used in practical filled rubbers for many years. Research studies in nanoscale inclusions include the following. Polymer matrix nanocomposites may exhibit an increase or reduction in glass-transition temperature (T_g) in comparison to the pure polymer [22] [23]. Above T_g the polymer is comparatively compliant and extensible; below, it is much stiffer. In silica - polystyrene nanocomposites the T_g decreases by 11 K with increasing filler content. In thin polymer layers a gradient in T_g extends tens of nanometers into the film [24]. In PMMA (polymethyl methacrylate polymer) with aluminum oxide inclusions 39 nm in size, T_g decreases by about 25°C for filler concentration greater than 0.5% [27]. The underlying cause of the T_g suppression in constrained geometries remains a subject of inquiry. Interfaces and surfaces also slow the rate of physical aging, also known as structural relaxation, in polymethyl methacrylate (PMMA) [28]. The reduction of aging rate is a factor of 2 near a free surface and a factor of 15 at an interface with silica.

As for dental composites (§5.2.1) that contain mineral inclusions in a polymer matrix, several nano-filled restorative materials have been produced by various manufacturers with a filler size ranging from 5 to 100 nm. Nano-filled composite resins are claimed to have improved properties in both aesthetics and mechanical performance. Superior mechanical properties such as high flexural strength, low abrasion, low polymerization shrinkage and resistance to fracture are attributed to the high filler concentration of these materials, considered to be enabled by the small size the particulate inclusions possess.

5.2.5 Self healing polymers

The motivation for self healing materials is as follows [25]. Structural materials including polymers are susceptible to damage in the form of cracks. These cracks can form on the surface or deep within the solid where they are difficult to detect and impossible to repair. Cracking damage due to overload or fatigue also leads to mechanical degradation of polymer composites. In microelectronic polymeric components such damage can also lead to electrical failure.

A self healing polymer material [25] contains nearly spherical capsules (0.05 to 0.2 mm diameter) of a monomer-based liquid healing agent. Growth of a crack breaks some of the capsules, releasing the healing agent. Polymerization of the healing agent is then triggered by contact with an embedded catalyst; this bonds the crack faces. Most of the initial toughness is recovered via this healing process. Such materials have the potential to increase the reliability and service life of thermosetting polymers. More recently, self healing polymers have been developed based on micro-vascular networks of channels that contain healing agent, §4.4.

5.3 Fibrous polymer matrix solids

5.3.1 Why fibers?

Fibers used in composites exhibit stiffness and strength that are highly favorable compared with strength for bulk engineering materials. Slender fibers are strong because they can be prepared with very few defects such as micro-cracks. Defects act as stress concentrators and so reduce the strength.

Representative properties of fibers are given in Table 5.1. Steel used for fibers may be of a higher strength variety than mild structural steel; toughness is not as important for a fiber as in a bulk material because fibers are too thin to support propagating cracks.

Table 5.1: Fiber materials: density ρ, Young's modulus E and tensile strength σ^{ult}. Specific modulus E/ρ and specific strength σ^{ult}/ρ are in units of GPa/g/cm^3. Adapted from [6].

Material	ρ (g/cm^3)	E (GPa)	σ^{ult} (GPa)	E/ρ	σ^{ult}/ρ
Steel	7.8	210	0.34-2.1	27	0.043-0.27
Boron	2.63	385	2.8	146	1.1
Graphite, hi E	1.9	390	2.1	205	1.1
Graphite, hi σ^{ult}	1.9	240	2.5	126	1.3
E-Glass	2.54	72.4	3.5	28.5	1.38
Kevlar 49	1.50	130	2.8	87	1.87

5.3.2 Unidirectional fibrous composites

In fibrous polymer matrix composites, fibers made of a stiff, strong material are embedded in a polymer matrix. Fully aligned fibrous composites attain the Voigt upper bound $E_c = E_1V_1 + E_2V_2$ on stiffness as discussed in §2.3.1. Such composites are considerably lighter in weight than structural metals of comparable stiffness and strength. For example [6], a graphite-epoxy composite with 63% by volume of aligned fibers has a longitudinal Young's modulus E_L of about 160 GPa compared with 200 GPa for structural steel, a tensile strength σ_{ult} of 1.7 GPa compared with 0.5 GPa for structural steel, and a density of 1.6 g/cm^3, compared with 7.8 g/cm^3 for structural steel. These materials are highly anisotropic. The transverse properties of such a unidirectional composite are considerably less than the longitudinal properties: $E_T = 11$ GPa, $\sigma_{ult} = 42$ MPa. Also the shear modulus in-plane is $G = 6.4$ GPa and the Poisson's ratio for load in the longitudinal direction is $\nu_{LT} = 0.38$. Unidirectional fibrous materials are used in applications such as struts in which the stress is always along the long axis.

The Halpin-Tsai equations [6] provide reasonable approximations for the transverse modulus E_T of fibrous composites in terms of the matrix modulus E_m, the fiber modulus E_f, the volume fraction V_1 and the fiber shape.

$$\frac{E_T}{E_m} = \frac{1 + \xi\eta V_1}{1 - \eta V_1}, \tag{5.1}$$

in which

$$\eta = \frac{E_f/E_m - 1}{E_f/E_m + \xi}, \tag{5.2}$$

with ξ as a geometric parameter. For fibers of circular or square cross-section, $\xi = 2$. For rectangular fibers of dimensions a and b, $\xi = 2(a/b)$. As expected, the transverse modulus from the Halpin-Tsai equations is greater than the Reuss lower bound. If the fibers are stiff, the transverse modulus is much lower than the longitudinal modulus.

Stress concentration factor

Fibrous composites can be highly anisotropic. For anisotropic solids the stress concentration factor around holes can become large in comparison with the value in isotropic solids as discussed in §3.4.

5.3.3 Laminates

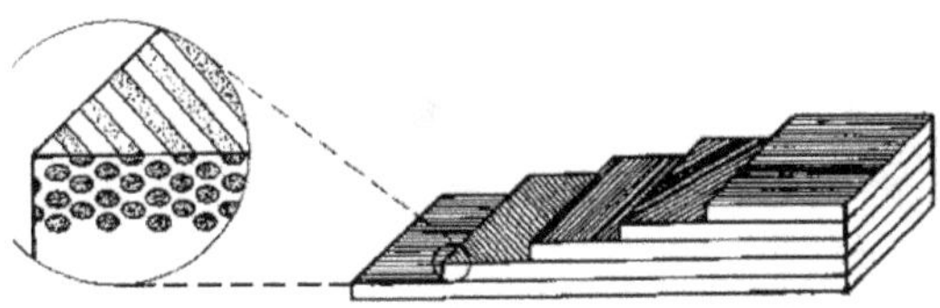

Figure 5.3: Hierarchical structure of a practical fibrous laminate [3], with permission.

Figure 5.4: Angle ply graphite epoxy fibrous composite, cross-section image [26], with permission. Scale bar, 100μm.

Lamination is used to control the anisotropy of a fibrous composite. The rationale is that if stiffness and strength are needed in directions other than the longitudinal direction, unidirectional composites are inadequate because they are too compliant and weak in shear and in tension in the transverse direction. Therefore cross-ply or angle ply laminated microstructures are designed and fabricated. Each lamina may contain fibers in a different direction. Each layer or lamina contains unidirectional fibers.

The number of layers, their angle and their organization are specified by a lamination code. The subscripts in the lamination code represent the number of plies or layers, and s represents a laminate symmetric about its mid-plane; the quantities in the brackets represent the angle of each ply with respect to a given direction. For example a unidirectional laminate of n layers is denoted $[0]_n$, with the zero representing a zero degree angle. Such a laminate is highly anisotropic. Fiber-dominated graphite-epoxy lay-ups such as $[0]_{48}$ and $[0/45/0/-45]_{6s}$ are stiff. Matrix-dominated lay-ups such as $[90]_{48}$ and $[90/-45/90/-45]_{6s}$ are more compliant. The hierarchical structure of a practical anisotropic fibrous material shown in Figure 5.3 differs from the hierarchical structures used in §2.3.5 to attain upper bounds for isotropic composites.

Laminate analysis

Laminates of fibrous layers (Figure 5.4), are analyzed as follows. Only a bare bones summary is presented here; the detailed analysis presented in many treatments (e.g. [6]) can constitute a large fraction of the content of a course.

To calculate the response of the laminate to load, one begins with the properties of an individual lamina. The properties of each lamina are considered in a two-dimensional form. The Q matrix in Equation 5.3 is the elastic modulus in two dimensions.

$$\begin{pmatrix} \sigma_1 \\ \sigma_2 \\ \tau_{12} \end{pmatrix} = \begin{pmatrix} Q_{11} & Q_{12} & 0 \\ Q_{12} & Q_{22} & 0 \\ 0 & 0 & Q_{66} \end{pmatrix} \begin{pmatrix} \epsilon_1 \\ \epsilon_2 \\ \gamma_{12} \end{pmatrix} \tag{5.3}$$

As with the reduced notation C_{ij} matrix in three dimensions, the Q_{ij} matrix is not a tensor and does not transform as a tensor, therefore a matrix for rotation must be derived separately. Similarly the stress and strain are displayed as column vectors but they do not transform under rotation as column vectors. Calculation of rotated Q_{ij} matrices, called $\bar{Q_{ij}}$, is done to determine the properties of laminae that are incorporated into the laminate at different angles. The effective elastic modulus A_{ij} of the laminate is calculated via a thickness-weighted summation of stiffness matrices of the corresponding layers, giving rise to an extensional stiffness matrix of the laminate as expressed in Equation 5.4. The analysis is similar to that of the Voigt composite. Each constituent matrix is based on the anisotropic elastic properties of the corresponding lamina. There are assumed to be n laminae. The quantity h_k represents the position in depth of lamina number k.

$$A_{ij} = \sum_{k=1}^{n} (\bar{Q_{ij}})_k (h_k - h_{k-1}) \tag{5.4}$$

Further stiffness matrices are used to account for effects of dependence of lamina properties on position.

$$B_{ij} = \frac{1}{2} \sum_{k=1}^{n} (\bar{Q_{ij}})_k (h_k^2 - h_{k-1}^2) \tag{5.5}$$

$$D_{ij} = \frac{1}{3} \sum_{k=1}^{n} (\bar{Q_{ij}})_k (h_k^3 - h_{k-1}^3) \tag{5.6}$$

Define $\mathbf{N}$ as the set of components of normal and shear force per unit length upon the edges of the laminate. $\mathbf{N}$ is expressed in terms of integrals of the stress components in each lamina over the total thickness h of the laminate.

$$\mathbf{N} = \begin{pmatrix} \int \sigma_x dz \\ \int \sigma_y dz \\ \int \tau_{xy} dz \end{pmatrix} \tag{5.7}$$

Similarly define $\mathbf{M}$ as the set of components of bending and twisting moment per unit length upon the edges of the laminate, expressed in terms of the stress distribution.

$$\mathbf{M} = \begin{pmatrix} \int \sigma_x z dz \\ \int \sigma_y z dz \\ \int \tau_{xy} z dz \end{pmatrix} \tag{5.8}$$

Then the force and moment per length on the edges are related to the deformation by the following.

$$\mathbf{N} = \mathbf{A}\epsilon + \mathbf{B}\kappa \tag{5.9}$$

in which ϵ is the strain (three components) in Equation 5.3 and κ is the curvature (three components) and $\mathbf{A}$ is the stiffness matrix in Equation 5.4 and $\mathbf{B}$ is the stiffness matrix in Equation 5.5.

Similarly,

$$\mathbf{M} = \mathbf{B}\epsilon + \mathbf{D}\kappa \tag{5.10}$$

in which **B** is the stiffness matrix in Equation 5.5 and **D** is the stiffness matrix in Equation 5.6.

This formulation accounts for the stiffness of the laminate in extension or compression in different directions or in shear; also the coupling between extension, bending and twisting which results from laminae at oblique angles that are displaced from the neutral axis. Specifically the **B** matrix quantifies coupling between extension and bending. This coupling does not arise from anisotropy in the laminae; it will occur in laminates of isotropic laminae of different stiffness. Instead, the bend-stretch coupling arises from the asymmetry in the stacking of the laminae. Similarly the **D** matrix incorporates rigidity in bending and twisting as well as coupling between bending and twisting.

Quasi-isotropic laminates

It is possible to obtain isotropic in-plane properties by proper choice of the stacking sequence. Quasi-isotropic laminates simplify some aspects of design; they are appropriate if the load may come from any direction in-plane. The requirement is that the extension stiffness matrix **A** is isotropic in-plane. Such a laminate requires three or more laminae; the laminae have identical stiffness matrices **Q**, and the laminae are oriented at equal angles [6]. This laminate is isotropic in-plane but the properties in the direction orthogonal to the laminate plane will be different from the properties in-plane.

The requirements for the components of **A** are $A_{11} = A_{22}$, $A_{11} - A_{12} = 2A_{66}$ and $A_{16} = A_{26} = 0$.

Cross-ply laminates; stacking sequence

Representative properties of practical cross-ply laminates in comparison with steel and aluminum alloy are given in Table 5.2. If the material is to be used in bending, the figure of merit is $\frac{E}{\rho^2}$ for rod bending or $\frac{E}{\rho^3}$ for plate bending as analyzed in §1.3; for bending the composite materials are more advantageous compared with steel or aluminum.

Cross-ply laminates have cubic symmetry in-plane; the orthogonal direction has different properties, so in three dimensions they are orthotropic.

Laminates behave differently from homogeneous materials in bending or torsion. For example, a [0/90/90/0] laminate and a [90/0/0/90] laminate are identical in tension but [0/90/90/0] is stiffer in bending because the longitudinal plies are further from the neutral axis. Also, a laminate such as [45/0/0/ − 45] can exhibit stretch twist coupling [29]. There are few laminae so this is structurally chiral rather than a chiral effective con-

Table 5.2: Cross ply fiber composites glass-epoxy, carbon-epoxy and boron-epoxy in comparison with bulk engineering materials: density ρ, fiber volume fraction V_f in percent, Young's modulus E and tensile strength σ^{ult}. Specific modulus E/ρ and specific strength σ^{ult}/ρ are in units of GPa/g/cm^3. Adapted from [6].

Material	ρ (g/cm^3)	V_f	E (GPa)	σ^{ult} (GPa)	E/ρ	σ^{ult}/ρ
Mild steel	7.8		210	0.45-0.83	27	0.058-0.11
Al 6061	2.7		69	0.26	26	0.096
Glass-ep	1.97	57	22	0.57	11	0.26
Carbon-ep	1.54	58	83	0.38	54	0.24
Boron-ep	2.0	60	106	0.38	53	0.19

tinuum. The role of controlled anisotropy is illustrated in the experimental X-29 military aircraft. The forward swept wings (Figure 5.5) allow extremely high maneuverability. However such wings are problematical for stability [30], because bending of the wing during maneuvers will cause twist that increases the wing angle of attack. This amplifies the acceleration of the maneuver, causing instability. The effect is neutralized via designed bend-twist coupling in the composite comprising the wings.

Figure 5.5: Grumman X-29 aircraft with forward swept wings [30], with permission.

Anisotropy in thermal expansion of unidirectional laminae (described above) gives rise to residual stress in laminates, which are prepared at elevated temperature and used at a lower temperature.

5.3.4 Nano-tubes as fibers

Carbon nano-tubes offer impressive modulus (1000 GPa) and strength (200 GPa) so they have been considered to be promising fiber materials. The high properties result from the stiff and strong bonds between carbon atoms and the fact that the structure of atoms in the tubes is free of defects. Composites based on them have not performed nearly as well as expected.

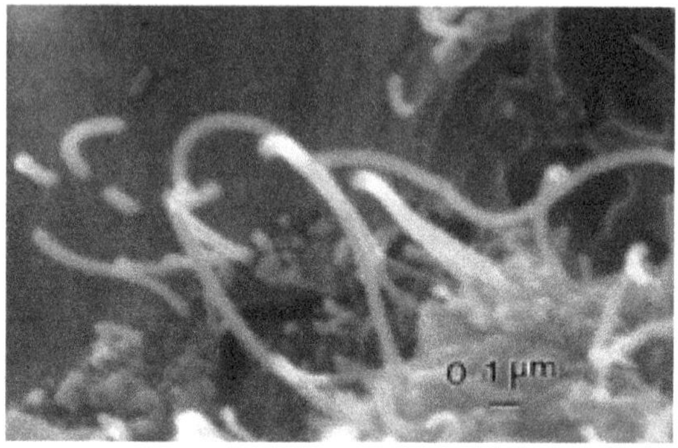

Figure 5.6: Curled nano-tube fibers [33] with permission. Scale bar, 0.1 μm.

During growth, the tubes may assemble as multi-walled concentric tubes or as bundles or ropes of single walled tubes [31]. These forms are pertinent to the tube performance in composites. Nano-tubes are typically randomly dispersed in a solvent or polymer melt via ultrasound or shear mixing, then processed further to make the composite. Such composites are often weaker than the matrix itself; if they are stronger it is usually not by much [32]. Many authors report that higher concentrations of tubes give rise to poorer performance than lower concentrations, in contrast to expectations from theory.

Suggested reasons for the lack of performance of these composites are as follows. Multi-walled tubes exhibit slip between the layers. This is perhaps not surprising in view of the fact that graphite has such weak interaction orthogonal to its planes that it is used as a lubricant [32]. Tubes in bundles slip with respect to each other. During processing, the tubes tend to twist and curl, reducing any reinforcement effect of the tubes (Figure 5.6). Moreover, the interface with the matrix tends to slip. Strain in the tubes has been measured as the composite is deformed, using Raman spectroscopy [33]. Load transfer from matrix to tube was better in compression than in tension. One can also assemble membranes layer by layer to obtain better performance but that is not practical for structural materials.

More success has been reported with using nano-tubes for enhancement of other physical properties such as electrical conductivity and thermal conductivity in polymer matrix composites [32].

Nanotubes also are only available in considerably shorter lengths (at most a fraction of a millimeter to as much as a millimeter or two) than commonly used graphite or boron fibers; this creates problems with stress transfer [34]. Weak interaction with the matrix is solved in graphite fibers by chemical modification, but nanotubes are one atom thick, so attempts to chemically modify the tubes substantially compromises their physical properties. Graphene, which is a layer of carbon atoms one atom thick, has also been explored as a composite constituent. Its challenges are similar to those encountered with carbon nano-tubes.

5.3.5 Effects of moisture

Composite materials based on epoxy resins have been used in a variety of applications in the aerospace industry. These materials exhibit excellent mechanical properties and low density. However, the epoxy resins used in such systems usually absorb water or moisture. Variations in temperature can cause a substantial degradation of the resin's mechanical properties and these deleterious effects can be enhanced when the resin is also exposed to moisture. Exposure to fluctuating temperature and moisture is inevitable in aerospace applications. Moisture alters the glass transition temperature in both neat resins and fibrous composites [35]. The effect of moisture and temperature on the glass transition temperature of the matrix manifests itself in a reduction of stiffness and an increase in the mechanical damping of the composite [36].

5.3.6 Damage

Damage in composite laminates or sandwich panels is usually not as obvious as it is in metals. For example, overload of a metal plate will result in a visible dent. A composite subjected to a similar overload may spring back to its original shape in which damage is not obvious, yet fiber breakage and delamination has occurred, resulting in loss of ability to carry load. Crushing of the core will weaken the panel even if the face sheets are not obviously damaged. This aspect of composites can be problematical for aircraft. Damage occurs in commercial airliners that are bumped by service vehicles or by debris on the runway thrown up during takeoff or landing. Non-destructive testing including acoustic methods [37] [38] are therefore used to detect damage and to evaluate the structural integrity of the aircraft. As for the design of composite structures, it is desirable in this context that they be damage resistant as well as damage tolerant, so that after damage they retain the ability to support as much load as possible.

5.3.7 Making fibrous composites

Laminates can be made by manually laying up layers of fibrous material on a mold. Such layers are commercially available as tape; the fibers are pre-impregnated with matrix, giving rise to the name prepreg. The laminate may be compressed during curing via a flexible bag activated by compressed air (pressure bagging) or by pumping air from under the bag (vacuum bagging). Either process reduces porosity and improves quality. If chopped fibers are used, a mixture of fibers, resin and catalyst may be sprayed onto the mold, where the resin polymerizes. Laminates are also made by filament winding [6]. In this method the desired part or a hollow template or mandrel is rotated around an axis and filaments of fiber are progressively wound onto it. The fiber passes through a resin bath before winding. The fiber angle is controlled by the motion of the source of filaments.

5.4 Platelet reinforcement

Figure 5.7: Platelet reinforced aluminum alloy [21], with permission; original magnification, about 2000.

Platelets as reinforcement offer the advantage that they provide more stiffening effect than fibers or particles per unit volume as discussed in §2.4.3. Platelets are used as reinforcing constituents in ceramic, glass and metal matrices. Commercially available alumina and SiC platelets of approximately hexagonal shape and of axial ratios between 0.05 and 0.5 have been used. Platelets of these materials can be single crystal. Platelets have the advantage of good thermal stability and freedom from defects, low cost, environmental safety (compared with whiskers) [8]. Platelet inclusions tend to improve toughness. The effect on strength may be an increase or a decrease depending on the specific matrix. An aluminum alloy with SiC platelet inclusions (Figure 5.7) is made stronger by virtue of the increase in dislocation density that results from thermal expansion mismatch.

As for biological materials, mother-of-pearl (nacre) [39] is a platelet-reinforced composite, highly filled with calcium carbonate (aragonite). The inclusions are packed in a brick-like pattern with a thin layer of organic material between them. Electron microscopy reveals that the ductility of wet nacre is caused by cohesive fracture along platelet lamellae at right angles to the main crack. The matrix appears to be well bonded to the lamellae, enabling the matrix to be stretched across the delamination cracks without breaking, thereby sustaining a force across a wider crack. The importance of nacre to the mollusk is as follows. The catastrophic failure of the shell enables a predator to eat the mollusk. Predators are able to store strain energy which can be fused to strike the mollusk in an attempt to break the shell.

Materials inspired by nacre have long been studied in an effort to develop tough ceramic materials. Recently a composite with structure resembling the brick and mortar platelet structure of nacre was developed [40] to provide impact resistance combined with a transparency usually associated with glass, which is brittle.

5.5 Metal matrix composites

5.5.1 Particulate metal matrix composite stiffness and strength

Ceramic particles in a metal matrix are used to improve the strength of the metal. The improvement in strength is attributed to a higher dislocation density that arises from the mismatch in thermal expansion between matrix and inclusion. The strength improvement is usually accompanied by a reduction in ductility, particularly for high concentrations of inclusions. Particulate inclusions of spherical or near spherical shape do not have as much stiffening effect per unit volume as fibers but they are less expensive than fibers, so they are often used. Metal-matrix composites [41] include aluminum stiffened with inclusions of alumina (Al_2O_3) particles, and aluminum reinforced with graphite, boron, or silicon carbide. They are of particular interest in applications involving high temperature, for which a polymer matrix composite would be inappropriate. Such metal matrix composites exhibit increased stiffness, strength and wear resistance. For example [8], aluminum with 15% SiC particles by volume increases in modulus from 70 GPa to 98 GPa; 40% SiC by volume increases the modulus to about 148 GPa. Aluminum with short fiber or platelet inclusions has a greater yield strength than the matrix metal [42]. Some of the strength increase is attributed to dislocations in the matrix resulting from misfit strain due to the difference in thermal expansion between matrix and inclusion as the metal is cooled. A modified shear lag approach incorporating tensile transfer of load is helpful for understanding the increase in strength. Too high a concentration of inclusions reduces the ductility [43].

5.5.2 Nano-size particle inclusions in metal

Inclusions on the nanoscale help to achieve improved strength and maintain ductility [43]. The rationale is that larger size particle inclusions can have the effect of reducing toughness. These fine particles tend to clump together more than larger ones. The process of mixing of nano-sized ceramic particles in a metal matrix can be lengthy and expensive; the use of high intensity ultrasound to achieve mixing offers advantages [44]. Magnesium is less dense than aluminum. Magnesium based composites with nanoscale inclusions have been made [45]. Composites with a relatively high concentration (14 per cent by volume) of silicon carbide nanoparticles in magnesium were made through a self-stabilization mechanism in molten metal. Strength, stiffness, and plasticity were simultaneously improved.

5.5.3 Fiber inclusions in metal

Continuous SiC fibers have been used in an aluminum matrix. Mechanical properties [8] of composite with 47% SiC fibers by volume include longitudinal modulus 204 GPa, transverse modulus 118 GPa, longitudinal strength 1.5 GPa, transverse strength 86 MPa. Fibers of boron and of aluminum oxide have also been used. Matrices of titanium, magnesium and copper have also been used. The strength of these composites is maintained to higher temperatures than that of aluminum alone. The mismatch of thermal expansion between fiber and matrix generates dislocations during cooling. The matrix undergoes plastic deformation and strain hardening. So the matrix properties in the composite will differ from properties measured for a homogeneous specimen of matrix.

5.6 Composites with renewable constituents

The use of natural fibers in composites can be traced to the use of straw in bricks in ancient times [46]. More recently, cellulose fibers were incorporated in phenolic plastics in 1908 and fenders made with soy protein bio-plastic were used in Ford automobiles in 1941 [47]. The focus in modern natural fiber composites is not extreme mechanical performance but rather cost savings, recycling of materials that would otherwise be wasted, energy security, and environmental concerns [47].

Commercially available wood-plastic composites contain wood particles, typically less than 250 μm in width and less than 2 mm long, embedded in a polymer matrix [48] [49]. Polyethylene is the most common polymer used. The wood component has a relatively high strength to weight ratio, low density, and is renewable; it also can reduce the amount of polymer needed. The wood component can include byproducts such as residue from sawmills, wood from demolition of wood frame buildings, or trimmings from logging. The wood component is much cheaper than the plastic that it replaces in the composite. The wood particles typically contain bundles of short fibers rather than individual fibers. The polymer constituent renders these materials comparatively resistant to degradation by moisture and fungus. To make wood-plastic composites, the wood inclusions are dispersed in liquid polymer and can then be extruded to produce a part or made into pellets for further processing [48]. The increase of inclusion concentration will increase the Young's modulus of the composite above that of the polymer matrix but may decrease the strength, particularly for higher concentrations of inclusions [49]. These composites exhibit much more creep than does wood: up to a factor of 3 to 8 reduction in effective shear stiffness over time periods from about 30 seconds to one year [48]. Overall the mechanical properties are more similar to those of the polymer matrix than to those of wood; the composites have a lower strength than wood, and they experience more time and temperature-dependent behavior. Wood-plastic composites are used as outdoor decking material in residential construction; in park benches; also in house siding and in automotive parts.

5.7 Thermoelastic composites

The thermal expansion of composites is calculated using the same methods as those used for elastic moduli.

5.7.1 Thermal expansion, Voigt

Determine the thermal expansion α_c of a composite that obeys the Voigt model.

Solution

Use Hooke's law in one dimension for phases 1 and 2.

$\epsilon_1 = \frac{\sigma_1}{E_1} + \alpha_1 \Delta T$, $\epsilon_2 = \frac{\sigma_2}{E_2} + \alpha_2 \Delta T$.

For the Voigt condition $\epsilon_1 = \epsilon_2$. Solve for stress,

$\sigma_1 = E_1 \epsilon - \alpha_1 E_1 \Delta T$, $\sigma_2 = E_2 \epsilon - \alpha_2 E_2 \Delta T$.

Moreover the force applied to the composite is the sum of the forces supplied by each constituent. $F_c = \sigma_1 A_1 + \sigma_2 A_2$, with A as cross-section area. So the stress on the composite is $\sigma_c = \sigma_1 V_1 + \sigma_2 V_2$, with V as volume fraction. The volume fraction for this geometry is the ratio of constituent cross-section area to total cross-section area.

Combining,

$\sigma_c = E_1 V_1 \epsilon + E_2 V_2 \epsilon - \alpha_1 E_1 V_1 \Delta T - \alpha_2 E_2 V_2 \Delta T$.

For free expansion, $\sigma_c = 0$. Divide the above by strain, keeping in mind the definition of thermal expansion coefficient.

$E_1 V_1 + E_2 V_2 = \alpha_1 E_1 V_1 \Delta T/\epsilon + \alpha_2 E_2 V_2 \Delta T/\epsilon$. The left side is E_c so $E_c = \frac{\Delta T}{\epsilon}(\alpha_1 E_1 V_1 + \alpha_2 E_2 V_2)$.

Finally, $\alpha_c = \frac{\epsilon}{\Delta T} = \frac{(\alpha_1 E_1 V_1 + \alpha_2 E_2 V_2)}{E_c}$.

The one dimensional simplification neglects the effects of the Poisson's ratio of the constituents. The result applies to a unidirectional fibrous composite in the longitudinal direction.

5.7.2 Unidirectional composites: thermal expansion

The thermal expansion of unidirectional fibrous composite is anisotropic [6]; transverse expansion α_T is considerably greater than longitudinal expansion. The longitudinal expansion is

$$\alpha_L = (\alpha_f E_f V_f + \alpha_m E_m V_m)\frac{1}{E_L}$$

in which f refers to fiber and m refers to matrix. The transverse expansion α_T is given in approximate form for fiber volume fractions greater than 0.25,

$$\alpha_T = \alpha_f V_f + \alpha_m V_m (1 + \nu_m)$$

with ν_m as the Poisson's ratio of the matrix. These expansions differ from those based on elementary Voigt or Reuss analyses because Poisson's ratio is considered. The low expansion in the longitudinal direction can be useful in applications for which dimensional stability under temperature change is required.

As for representative thermal expansion coefficients [6], unidirectional graphite-epoxy composite has a transverse expansion expansion $\alpha_T = 20.2 \times 10^{-6}/°\text{C}$ and a longitudinal expansion $\alpha_L = 0.045 \times 10^{-6}/°\text{C}$. Glass-epoxy composite has a transverse expansion expansion $\alpha_T = 36 \times 10^{-6}/°\text{C}$ and a longitudinal expansion $\alpha_L = 5.4 \times 10^{-6}/°\text{C}$. So, the thermal expansion manifests considerable anisotropy in these unidirectional fibrous composites.

5.7.3 Thermal benders

Analysis of a bi-material bender strip [50] uses the usual elementary bending assumptions of a slender bar subject to small deflection, and there is

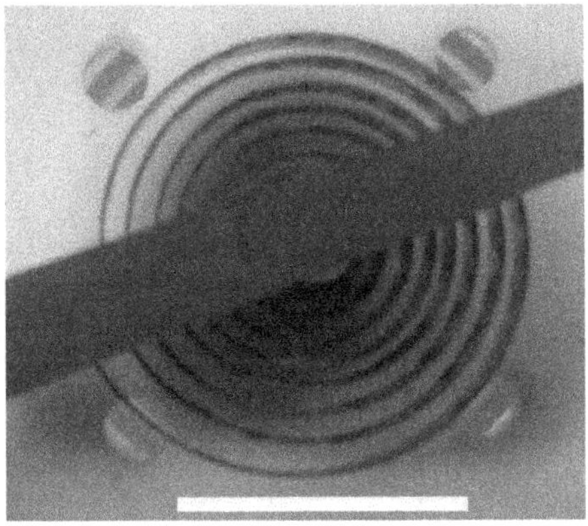

Figure 5.8: Bimetallic spiral. Scale bar, 10 mm.

no slip between the constituent materials. Given the elastic modulus and thermal expansion of each constituent, E_1 and α_1 for phase 1 and E_2 and α_2 for phase 2 respectively; also the thickness a_1 and a_2 of each layer in the bi-material strip, then the curvature resulting from a temperature change δT is [50],

$$\rho^{-1} = \frac{(\alpha_2 - \alpha_1)\delta T}{\frac{h}{2} + \frac{2(E_1 I_1 + E_2 I_2)}{h}[\frac{1}{E_1 a_1} + \frac{1}{E_2 a_2}]} \tag{5.11}$$

in which $I_1 = \frac{a_1^3}{12}$ and $I_2 = \frac{a_2^3}{12}$ are the section moments of inertia of the layers and $h = a_1 + a_2$ is the total thickness; ρ is the radius of curvature. If the thicknesses of each layer are identical, considerable simplification occurs as follows. The curvature is then not very sensitive to modulus ratio, giving rise to a further simplification on the right.

$$\rho^{-1} = 24\frac{(\alpha_2 - \alpha_1)\delta T}{h(14 + \frac{E_1}{E_2} + \frac{E_2}{E_1})} \approx \frac{3}{2}\frac{(\alpha_2 - \alpha_1)\delta T}{h}. \tag{5.12}$$

Thermal benders typically contain two metals with different thermal expansion and are called bimetallic strips. They may be shaped as flat strips, spirals, or helices, depending on the type of actuation desired. They are used for thermometers, thermostats, and actuators to control electric circuits based on temperature.

As an example consider a strip of thickness $h = 0.12$ mm, comprised of brass with $\alpha_1 = 19 \times 10^{-6}/°\text{C}$ and invar with $\alpha_2 = 1 \times 10^{-7}/\ °\text{C}$. Suppose the strip is initially straight. Determine the curvature and the radius of curvature for a temperature excursion of 10°C and for 82°C. Substituting, the curvature for 10°C is $\rho^{-1} = 2.36$ /m and the radius of curvature is $\rho = 0.42$ m. For 82°C the curvature is $\rho^{-1} = 19.3$ /m and the radius of curvature is $\rho = 0.052$ m. Such large deformation is useful in thermometers. To make the sensor more compact, a bimetallic spiral was used in a thermometer, Figure 5.8. The strip had the same thickness 0.12 mm and an outer diameter 16 mm at ambient temperature but unknown composition was used in a thermometer intended to measure over a range of 82°C. Over that range, the pointer connected to the helix in the thermometer rotates almost a full revolution.

By comparison a homogeneous strip of brass 100 mm long would expand (via the definition of thermal expansion) 1.9 μm for a temperature excursion of 10°C. Precision instruments in satellites and spacecraft are subject to much larger temperature excursions; the control of thermal expansion is important in such settings.

5.8 Piezoelectric composites

5.8.1 Piezoelectric composite structure and rationale

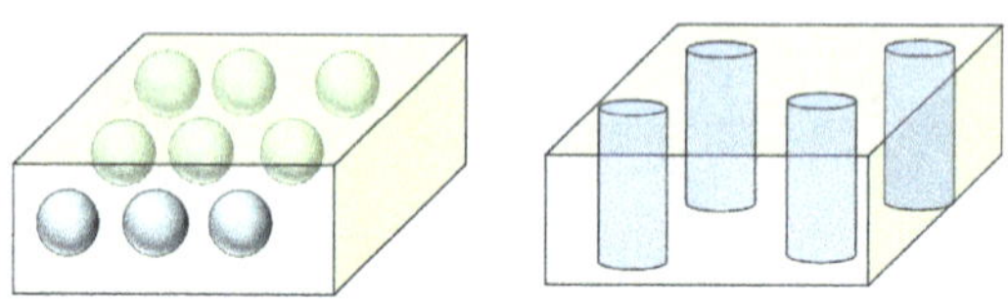

Figure 5.9: Piezoelectric composite structures; left, particulate; right, parallel fibers.

As for piezoelectric composite materials, structure can be particulate (Figure 5.9, left), fibrous, platelet or they may have a more complex geometry. Physical properties may change greatly depending on how the constituents are connected. Each phase in a composite may be self-connected in zero, one, two or three dimensions. The following terminology is used in the literature for the connectivity [51] of such materials. An unconnected checkerboard pattern is referred to as 0-0. A 2-2 piezoelectric composite contains alternate layers of piezoelectric and passive materials. Each phase is connected in two dimensions. A 1-3 composite consists of an array of piezoelectric pillars embedded in a passive dielectric matrix, usually polymer (Figure 5.9, right). Piezoelectric composites have advantages over homogeneous piezoelectric ceramic materials in that they may offer increased electromechanical coupling efficiency, also a decrease in specific acoustic impedance, and suppression of unwanted modes of vibration.

Piezoelectric composites [52] can offer enhancements of some coefficients compared with single phase materials. For example a Voigt type laminate (electrically parallel) of piezoelectric layers can enhance the g coefficient. In particular, envisage a composite containing a volume fraction V^1 of parallel ceramic laminae or rods, phase 1, of piezoelectric sensitivity (in the full tensor notation) d^1_{333} in a matrix that is much more elastically compliant than the ceramic and has a much smaller dielectric constant than the ceramic. The phase and composite designations are set as superscripts because subscripts are used to indicate direction. The composite is then compressed by a stress σ^c_{33} in the direction of the piezoelectric rods. The stress in the rods is

$$\sigma^1_{33} = \sigma^c_{33}/V_1.$$

The charge density on the surface is proportional to the stress via $\mathcal{D}_3 = d_{333}\sigma_{33}$ because the boundary condition on the electric displacement $\mathcal{D}$ reveals the surface charge density. The cross-sectional area of ceramic is a factor V^1 of the total area; since the force is the same, the stress is increased by the same factor, so

$$d^c_{333} = d^1_{333}.$$

The relative dielectric permittivity of the composite is

$$k^c_3 = V^1 k^1_3$$

via addition of capacitances in parallel; the capacitance contribution of the matrix is neglected because its dielectric constant is assumed to be much less than that of the piezoelectric ceramic. Finally

$$g^c_{333} = d^c_{333}/\epsilon_0 k^c_3 = d^1_{333}/V^1 \epsilon_0 k^1_3 = g^1_{333}/V^1.$$

So by choosing V^1 small, g^c_{333} can be made large. A similar procedure can be used to obtain a general form for arbitrary elastic moduli and dielectric properties [52].

A set of rule-of-mixtures type equations [51] was provided to approximate the effective piezoelectric coefficients of composites. The charge coefficient d (charge density due to polarization divided by stress) is given by a rule of mixtures. By contrast, the voltage coefficient g (open-circuit electric field divided by stress) in a laminar composite of piezoelectric and compliant dielectric layers perpendicular to the electrodes can be substantially greater than that of the piezoelectric constituent alone.

Hydrostatic sensitivity is pertinent to hydrophones that detect changes in hydrostatic pressure under water. The hydrostatic sensitivity of PZT ceramics is low because for these materials $d_{33} \approx -d_{31}$ (in the reduced notation). Piezoelectric composites containing parallel rods of ceramic offer superior sensitivity to hydrostatic pressure changes in comparison with the ceramic from which the rods are made [52]. If the matrix is made of negative Poisson's ratio material (§6.6), the sensitivity is improved further [53]. Further optimization [58] gave rise to a design with a porous polymer matrix with a negative Poisson's ratio.

5.8.2 Piezoelectric Voigt composite

Consider the piezoelectric sensitivity d^c_{33} of a Voigt composite with laminae oriented in the z or 3 direction and the stress in the same directions. This is a parallel electrical connection. After [52],

$$d^c_{33} = \frac{V^1 d^1_{33} S^2_{33} + V^2 d^2_{33} S^1_{33}}{V^1 S^2_{33} + V^2 S^1_{33}}$$

in which S_{33} is the elastic compliance in the 3 direction and V^1 is the volume fraction of phase 1 and d^1_{33} is its piezoelectric sensitivity. Superscripts are used here to identify the phase or constituent because subscripts are needed to identify direction.

To obtain the Voigt d sensitivity, consider the elastic Voigt construction, §2.3.1. The stress on the composite is σ^c_{33}; the corresponding strain is

$$\epsilon^c_{33} = \frac{\sigma^c_{33}}{E^c_3} = \frac{\sigma^c_{33}}{(E^1_3 V^1 + E^2_3 V^2)}.$$

The stress in phase 1 is $\sigma^1_{33} = \epsilon^1_{33} E^1_3$, so the electric displacement in the 3 direction of phase 1 is

$$\mathcal{D}^1_3 = d^1_{33}\sigma^1_{33} = d^1_{33}\epsilon^1_{33}E^1_3;$$

similarly for phase 2. Because the electric displacement is determined by the surface charge density, by considering the average surface charge density of the composite,

$$\mathcal{D}^c_3 = \mathcal{D}^1_3 V^1 + \mathcal{D}^2_3 V^2.$$

But $d^c_{33} = \frac{\mathcal{D}^c_3}{\sigma^c_{33}}$, so

$$d^c_{33} = (d^1_{33}E^1_3 V^1 + d^2_{33}E^2_3 V^2)\frac{\epsilon^c}{\sigma^c} = (d^1_{33}E^1_3 V^1 + d^2_{33}E^2_3 V^2)\frac{1}{E^c_3}.$$

To obtain the form cited, recall for the Voigt laminate, $E^c = E^1V^1 + E^2V^2$, so

$$S^c_{33} = \frac{1}{E^c_3} = \frac{S^1_{33} S^2_{33}}{V^1 S^2_{33} + V^2 S^1_{33}}.$$

Finally,

$$d^c_{33} = \frac{V^1 d^1_{33} S^2_{33} + V^2 d^2_{33} S^1_{33}}{V^1 S^2_{33} + V^2 S^1_{33}}.$$

One may verify that the case considered above, $d^2_{33} = 0$ and $S^2_{33} >> S^1_{33}$, a polymer matrix, gives rise to $d^c_{33} \approx d^1_{33}$.

As for the Reuss laminate, the effective piezoelectric sensitivity depends on the dielectric permittivity of the constituents as well as their piezoelectric sensitivity. In the Reuss configuration, a thin layer of low permittivity substantially lowers the d coefficient but has little effect on the g coefficient.

5.8.3 Piezoelectric benders

A laminate can be made using two layers of piezoelectric material of dissimilar polarity, or one layer of piezoelectric material cemented to an inert material such as brass. These are called bender elements; if there are two piezoelectric layers it is called a bimorph. When a voltage is applied to such a laminate, one layer expands, and the other contracts or remains the same, giving rise to bending.

The deflection u per volt V for a single bimorph cantilever rib of length L and full thickness h containing two anti-parallel piezoelectric layers of sensitivity d_{31} and thickness $h/2$ is [54] [55]

$$\frac{u}{V} = -\frac{3}{2}d_{31}[\frac{L}{h}]^2. \tag{5.13}$$

Elastic structural compliance (displacement / force) is $u/F = \frac{2}{w}S_{11}[\frac{L}{h}]^3$ with w as the width and S_{11} as the material compliance, inverse of Young's modulus E. Further details on the piezoelectric bimorph are given in [56]. A bender in which one layer is piezoelectric and the other layer is not piezoelectric, is called a unimorph; the analysis is more complicated [57] than for a bimorph. In comparing the piezoelectric bender with the thermal bender, keep in mind the piezoelectric sensitivity d has units m/V in contrast to the thermal expansion α which is dimensionless.

Deformation in bending is larger than it is in axial expansion. Piezoelectric benders are available in strip form that can be used as rib elements; also in disk form, usable as sound emitters, as other actuators, or as sensors.

5.8.4 Piezoelectric composite uses and fabrication

Piezoelectric composites find use in electromechanical transducers and in transducers intended for "smart" materials. Piezoelectric composites are available with particles of lead titanate piezoelectric ceramic embedded in a polychloroprene rubber [59]; also in columnar structures.

Several piezoelectric composite fabrication methods are as follows. In the dice-and-fill method, [60], deep grooves are cut into a solid ceramic block and a polymer is cast into the grooves. Slicing off the ceramic base leaves the desired 1-3 piezo-composite disk. To obtain a rod in matrix composite, one aligns long thin cylinders of piezoelectric ceramic with ends in the holes of two end pieces. Then one casts a polymer between them. After curing, the block is sliced into discs [61]. This method is effective for making samples with rod diameters about 0.2 mm or more. To obtain finer structure, two methods have been suggested. Firstly, carbon fiber is woven into the desired structure by textile techniques; then the carbon structure is replicated with piezoelectric ceramic. Secondly, a lost wax method is used in which a complementary structure is formed in polymer. Then a ceramic paste is injected into this mold and fired. The plastic mold burns away during the ceramic firing, and a polymer is cast back into its place.

5.9 In situ composites

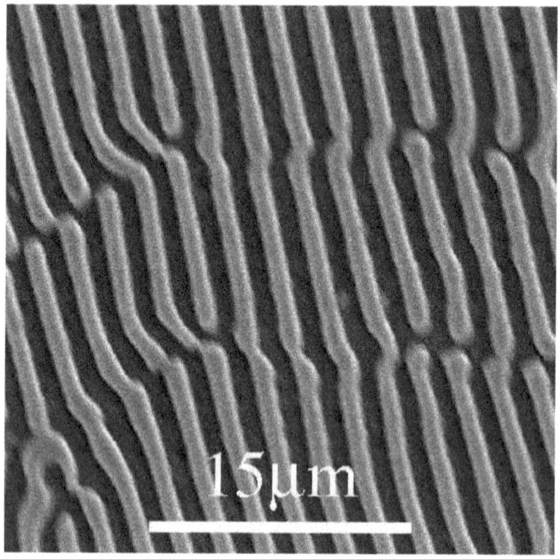

Figure 5.10: Structure of eutectic copper-aluminum alloy, [63], with permission. Scale bar: 15μm.

In-situ composites [62] are multiphase materials in which the reinforcing phase is synthesized within the matrix during composite fabrication. This is in contrast to ex-situ composites where the inclusion material is synthesized separately and then inserted into the matrix during a secondary process such as infiltration or powder processing.

Processes in which precipitation occurs in the liquid state include traditional and directionally solidified eutectics as well as rapid solidification in which a supersaturation is created by rapidly cooling a liquid with the reinforcement constituents already in solution. An image of a laminate structure formed during crystallization of a eutectic, Al 67, Cu 33 (wt%), eutectic alloy, is shown in Figure 5.10.

Reinforcements in discontinuously reinforced metallic or intermetallic matrix in-situ composites are on the order of 0.5-5 μm, and concentrations range from 0-50 vol.%. Advantages include smaller reinforcement particle size with higher strength, and better interface quality. Disadvantages include choice of in-situ reinforcements is limited to particles that are thermodynamically stable within a particular matrix. The size and shape of inclusion cannot be directly specified by the designer; rather, they are controlled by nucleation and growth processes.

5.10 Summary

Practical composite materials may contain particles, fibers, or platelets as inclusions. Particle composites include dental filler composites, tire rubber, and asphalt. Compliant particles are used to enhance toughness as in rubber modified polymers. Fibrous materials with short fibers offer stiffness and strength superior to that of the matrix combined with low cost. High-performance fibrous materials make use of stiff, strong fibers such as glass and graphite. Fibrous layers are organized into laminate structures to allow control of anisotropy.

Composite materials are also made to control or enhance the response of materials that admit coupled fields, as in thermo-elastic and piezoelectric materials.

Bibliography

[1] Z. Hashin and B. W. Rosen, The elastic moduli of fiber-reinforced materials, J. Appl. Mech., Trans. ASME, 31, 223-232, (1964).

[2] Z. Hashin, On elastic behaviour of fibre reinforced materials of arbitrary transverse phase geometry, J. Mech. Phys. Solids, 13, 119-134, (1965).

[3] R. S. Lakes, Materials with structural hierarchy, Nature, 361, 511-515, (1993).

[4] Y. Papadogianis, D. Papadogianis, Aristotle University of Thessaloniki, Greece, private communication.

[5] Cambridge University materials science micrograph archive, `https://www.doitpoms.ac.uk/miclib/full_record.php?id=558`

[6] B. D. Agarwal and L. J. Broutman, *Analysis and performance of fiber composites*, 2nd Ed. J. Wiley, New York, (1990).

[7] S. Ahmed, and F. R. Jones, A review of particulate reinforcement theories for polymer composites, J. Materials Sci., 25, 4933-4942, (1990).

[8] M. Schwartz, *Composite Materials*, Prentice Hall, Upper Saddle River, New Jersey (1997).

[9] R. N. J. Saal, Asphalt systems, ch. 9, in *Composite Materials*, ed. L. Holliday, Elsevier, Amsterdam, (1966).

[10] M. L. Cannon, Composite resins, in *Encyclopedia of Medical Devices and Instrumentation,* ed. J. G. Webster; J. Wiley, New York (1988).

[11] R. Craig, Chemistry, composition, and properties of composite resins, In Horn, H. (ed.) *Dental Clinics of North America*, Saunders, Philadelphia PA, (1981).

[12] Y. Papadogianis, D. B. Boyer, and R. S. Lakes, Creep of conventional and microfilled dental composites, J. Biomed. Materials Research, 18, 15-24, (1984).

[13] J. B. Park and R. S. Lakes *Biomaterials*, third edition, Springer, New York, (2007).

[14] 3M filtek supreme, 3M, Minneapolis, MN, (2019). `https://www.3m.com/`

[15] S. B. Mitra, D. Wu, B. N. Holmes, An application of nanotechnology in advanced dental materials, JADA 134, 1382-1390 (2003).

[16] R. N. Traxler, *Asphalt*, Reinhold, New York, (1961).

[17] R. S. Lakes, S. Kose, and H. Bahia, Analysis of high volume fraction irregular particulate damping composites, ASME Journal of Engineering Materials and Technology, 124, 174-178, (2002).

[18] H. Lanceley, J. Mann, G. Pogany, Thermoplastic polymer systems, in *Composite Materials*, ed. L. Holliday, Elsevier, Amsterdam, (1966).

[19] A. R. Payne, in *Rheology of Elastomers*, edited by P. Mason, N. Wookey, Pergamon, (p. 58), London, (1958).

[20] A. R. Payne, A note on the existence of a yield point in the dynamic modulus of loaded vulcanizates Journal of applied polymer science, 3, 127, (1960).

[21] R. J. Arsenault, The strengthening of aluminum alloy 6061 by fiber and platelet silicon carbide, Materials Science and Engineering, 64 171-181, (1984).

[22] A. Bansal, H. Yang, C. Li, K. Cho, B. C. Benicewicz, S. K. Kumar, L. S. Schadler, Quantitative equivalence between polymer nanocomposites and thin polymer films, Nature Materials 4, 693 - 698, (2005).

[23] A. M. Mayes, Nanocomposites: softer at the boundary, Nature Materials 4, 651 - 652, (2005).

[24] C. J. Ellison and J. M. Torkelson, The distribution of glass-transition temperatures in nanoscopically confined glass formers, Nature Materials 2, 695 - 700, (2003).

[25] S. R. White, N. R. Sottos, P. H. Geubelle, J. S. Moore, M. R. Kessler, S. R. Sriram, E. N. Brown and S. Viswanathan, Autonomic healing of polymer composites, Nature 409, 794-797, (2001).

[26] Liang Dong, Northeast University, China, with permission.

[27] B. J. Ash, L. S. Schadler, R. W. Siegel, Glass transition behavior of alumina / polymethylmethacrylate nanocomposites, Materials Lett. 55, 83-87, (2002).

[28] R. D. Priestley, C. J. Ellison, L. J. Broadbelt, and J. M. Torkelson, Structural relaxation of polymer glasses at surfaces, Interfaces and in between, Science 309, 456-459, (2005).

[29] R. M. Jones, *Mechanics of Composites*, Taylor and Francis, Abingdon, UK (1975).

[30] `https://en.wikipedia.org/wiki/Grumman_X-29`

[31] P. M. Ajayan, L. S. Schadler, C. Giannaris, Single walled carbon nanotube polymer composites: strength and weakness, Advanced Materials, 12 (10) 750 (2000).

[32] R. Andrews, M.C. Weisenberger, Carbon nanotube polymer composites, Current Opinion in Solid State and Materials Science 8, 31-37 (2004).

[33] L. S. Schadler, C. Giannaris, P. M. Ajayan, Load transfer in carbon nanotube composites, Appl. Phys. Lett. 73, 3842 (1998). Figure reproduced from this article with the permission of AIP Publishing.

[34] I. A. Kinloch, J. Suhr, J. Lou, R. J. Young, P. M. Ajayan, Composites with carbon nanotubes and graphene: an outlook, Science, 362(6414), 547-553 (2018).

[35] R. DeIasi, and J. B. Whiteside, Effect of moisture on epoxy resins and composites, Advanced composite materials- environmental effects, ed. J. R. Vinson, ASTM publication STP 658, Phila. PA, (1978).

[36] L. W. Rehfield, R. P. Briley, and S. Putter, Dynamic tests of graphite epoxy composites in *Hygrothermal environments*, ed. N. R. Adsit, ASTM publication 768, Phila. PA, (1982).

[37] L. P. Dickinson, N. H. Fletcher, Acoustic detection of invisible damage in aircraft composite panels, Applied Acoustics 70 (1) 110-119 (2009).

[38] P. Cawley, R. Adams, The mechanics of the coin-tap method of non-destructive testing. J Sound Vibr., 122, 299-316, (1988).

[39] A. P. Jackson, J. F. V. Vincent and R. M. Turner, The mechanical design of nacre, Proc. Royal Soc. Lond. B 22 415-440 (1988).

[40] Z. Yin, F. Hannard, and F. Barthelat, Impact resistant nacre like transparent materials, Science, 364, 1260-1263, (2019).

[41] M. Taya and R. J. Arsenault, *Metal Matrix Composites*, Pergamon, Oxford, (1989).

[42] V. C. Nardone and K. M. Prewo, On the strength of discontinuous silicon carbide reinforced aluminum composites, Scripta Metall. 20, 43-48 (1986).

[43] K. Akio, O. Atsushi, K. Toshiro, T. Hiroyuki, Fabrication process of metal produced by vortex method, J. Jpn. Inst. Light Met. 49 149-154 (1999).

[44] Y. Yang, J. Lan, X. Li, Study on bulk aluminum matrix nano-composite fabricated by ultrasonic dispersion of nano-sized SiC particles in molten aluminum alloy, Materials Science and Engineering A 380, 378-383 (2004).
[45] L. Y. Chen, J. Q. Xu, H. Choi, M. Pozuelo, X. Ma, S. Bhowmick, J. M. Yang, S. Mathaudhu, X. C. Li, Processing and properties of magnesium containing a dense uniform dispersion of nanoparticles, Nature, 528, 539-543 (2015).
[46] Exodus, 5.
[47] A. K. Mohanty, S. Vivekanandhan, J. Pin, M. Misra, Composites from renewable and sustainable resources: challenges and innovations, Science, 362, 536-542 (2018).
[48] S. Hamel, J. Hermanson, and S. Cramer. Mechanical and time-dependent behavior of wood-plastic composites subjected to bending, Journal of Thermoplastic Composite Materials, 28(5), 630-642 (2015).
[49] M. J. Schwarzkopf and M. D. Burnard, Wood-plastic composites- performance and environmental impact, in A. Kutnar and S. S. Muthu (eds.), *Environmental Impacts of Traditional and Innovative Forest-based Bioproducts*, Springer, Singapore, (2016).
[50] S. P. Timoshenko, Analysis of bi-metal thermostats, J. Optical Soc. America, 11, 233-355, (1925).
[51] D. P. Skinner, R. E. Newnham, and L. E. Cross, Connectivity and piezoelectric-pyroelectric composites, Mater. Res. Bull. 13, 599, (1978).
[52] R. E. Newnham, D. P. Skinner, L. E. Cross, Composite piezoelectric transducers, Materials in Engineering, 2 93-106 (1980).
[53] W. A. Smith, US Patent 5334903A - Composite piezoelectrics utilizing a negative Poisson ratio polymer, filed 12-04 (1992).
[54] J. G. Smits and S. I. Dalke, The constituent equations of piezoelectric bimorphs, IEEE Proc. 1989 Ultrasonics Symp. p. 781-784 (1989).
[55] J. G. Smits and S. I. Dalke and T. K. Cooney, The constituent equations of piezoelectric bimorphs, Sensors and Actuators A, 28, 41-61, (1991)
[56] M. R. Steel, F. Harrison, and P. G. Harper, The piezoelectric bimorph: an experimental and theoretical study of its quasistatic response, J. Physics D Appl. Phys. 11, 979, (1978).
[57] J. G. Smits and W. S. Choi, The constituent equations of piezoelectric heterogeneous bimorphs, IEEE Trans. Ultrason., Ferroelectrics and Frequency Control 38, 256-270 (1991).
[58] L. V. Gibiansky and S. Torquato, On the use of homogenization theory to design optimal piezocomposites for hydrophone applications, J. Mech. Phys. Solids, 45, 689-708 (1997).

[59] K. Rittenmyer, Temperature dependence of the electromechanical properties of 0-3 $PbTiO_3$ polymer piezoelectric composite materials, J. Acoust. Soc. Am., 96, 307-318, (1994).
[60] T. R. Gururaja, A. Safari, R. E. Newnham, and L. E. Cross, Piezoelectric ceramic-polymer composites for transducer applications, in *Electronic Ceramics*, ed. L. M. Levinson, pp. 92-128, Marcel Dekker, New York, (1987).
[61] W. A. Smith, The role of piezocomposites in ultrasonic transducers, IEEE 1989 ultrasonics symposium - 755-766 (1989).
[62] R. M. Aikin, Jr., The mechanical properties of in situ composites, JOM 35-39, (1997).
[63] Cambridge University materials science micrograph archive, `https://www.doitpoms.ac.uk/miclib/full_record.php?id=4`

Chapter 6

Cellular solids and lattices

6.1 Introduction

Cellular solids are composites in which one phase is empty space or a fluid such as water or air. Cellular solids include honeycombs, foams, and other porous materials [1]. Biological materials such as bone and wood have cellular structure as well as fibrous structure. Lattices are cellular solids in which the structure contains a solid phase which repeats in three dimensions. In this chapter, tilings / tessellations are considered first, then the mechanical properties of honeycombs (§6.3) and foams (§6.4) are analyzed. Lattices (§6.5) are then studied.

Use of void space in cellular solids allows one to attain negative and extreme physical properties. Extremal modulus and strength is achieved in hierarchical lattices that have structure within structure, §6.5.6. The tuning of Poisson's ratio to extreme values including negative Poisson's ratio is presented in §6.6. The tuning of thermal expansion to negative or extreme values is presented in §6.7.1. Tuning of the Hall effect is presented in §6.7.3. The control of waves including cloaking and wave absorption is presented in §6.8.

6.2 Tessellations

A cellular solid with a sufficiently regular structure can be represented as a tiling. A tiling, also called a tessellation, is a division of a plane or of three-dimensional space into geometrical figures [2]. In the plane, there are three regular polygons that can tile the plane: the square, the regular hexagon and the equilateral triangle (Figure 6.1). Regular hexagons are the most commonly used shape for honeycomb cells. One can also tile the plane with combinations of regular polygons [3] [4] [5]. For example, squares and

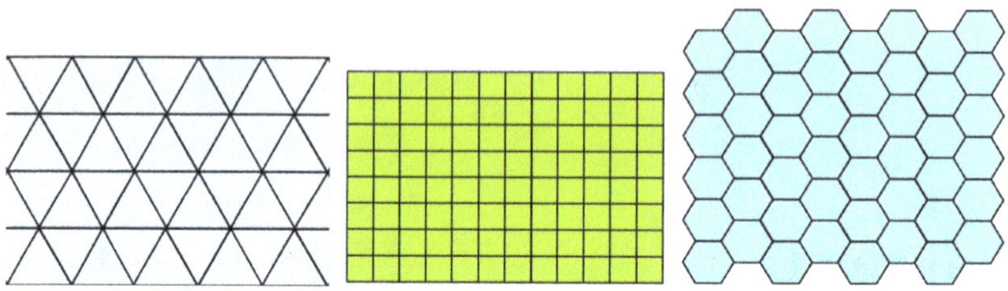

Figure 6.1: Tiling of the plane by regular polygons.

octagons tile the plane as shown in Figure 6.2; octagons can be chosen as regular.

Figure 6.2: Tiling of the plane by squares and octagons.

In three dimensions, the cube is the only Platonic solid that can tile 3D space [6]. Several semi-regular polyhedra can tile 3D space [5]. The Platonic solids have faces that are identical regular polygons. The semi-regular polyhedra have faces that are also regular polygons but they are not identical. The truncated octahedron or tetrakaidecahedron, for example, has square and regular hexagonal faces. It is an Archimedean solid which tiles space. It has been studied as a model for foam cells and as a cell for regular cellular lattices. Archimedean solids are convex polyhedra that have faces which are non-identical regular polygons meeting in identical vertices, and excluding the prisms and antiprisms. Their edges have identical vertices and equal length. The rhombic dodecahedron can also be packed to tile space. The rhombic faces are identical but are not regular polygons. These polyhedra that tile space are shown in Figure 6.3.

One may also tile space with combinations of polyhedral shapes. For example regular octahedra may be combined with regular tetrahedra giving

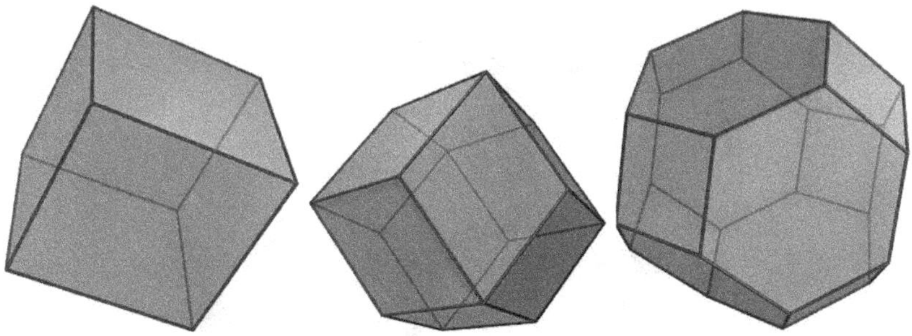

Figure 6.3: Solids that can tile space: cube [7], rhombic dodecahedron [8], truncated octahedron [9], with permission.

rise to the octet truss after Buckminster Fuller [10]. This structure was envisaged in 1967 in the context of large scale building construction; it has been studied recently (§6.5.1) in the context of lattices with cells on a mm or μm scale.

The above tessellations fill space with no gaps. If one allows gaps, one may also fill space with other regular figures. Dense packing of Platonic and Archimedean solids have been considered from a mathematical perspective [11] [12]. Some packing arrangements correspond to a regular array and are considered as Bravais lattices. The Bravais lattice is an array of points which when repeated can fill space. When the points represent atoms, ions or molecular chains, the lattice is considered to be a crystal. In the context of cellular solids, the points can represent nodes at which rib or plate elements intersect. The dense packing of solid figures has also been considered in the context of models for the structure of matter and in the study of granular materials.

6.3 Honeycomb

Honeycombs have periodic structures in two dimensions. They may be regarded as 2D lattices (§6.5). In a physical honeycomb the two-dimensional shape is projected into the orthogonal direction to generate the three-dimensional model of the physical structure.

A hexagonal cell honeycomb has cell wall lengths L and H in-plane. The depth in the out-of-plane (z) direction is B as shown in Figure 6.4. The angle θ and the ratio of L to H determine the shape of the hexagonal cell. The cell wall material (the solid phase) is assumed to have Young's modulus E_s and yield strength $\sigma_{y,s}$. Analyses of honeycomb and foam are based on those of Gibson and Ashby [1].

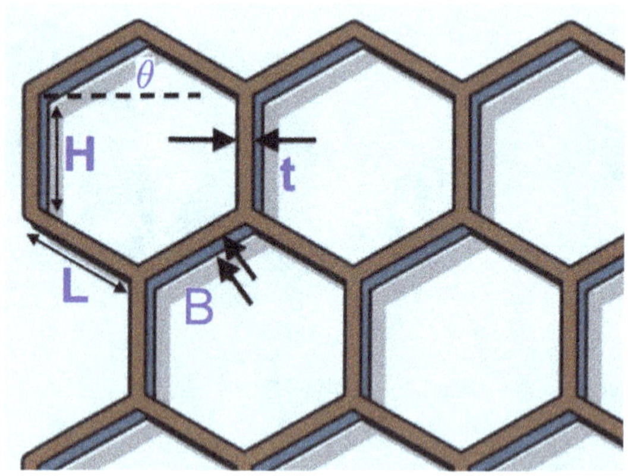

Figure 6.4: Hexagonal cell honeycomb diagram

Structural honeycomb (Figure 6.5) is typically used in light-weight sandwich structures (Figure 6.6) that are stiff and strong in bending. The sandwich typically contains a core of structural honeycomb or foam, and bonded face-sheets of graphite-epoxy composite. The density depends on

Figure 6.5: Hexagonal cell honeycomb core for sandwich panel structure. Scale bar: 10 mm.

aspect ratio for regular hexagons according to $\frac{\rho}{\rho_s} = C_1 \frac{t}{L}$ In which ρ_s is the density of the solid phase and ρ is the density of the cellular solid, with $C_1 = 2/\sqrt{3}$.

Honeycomb consists of plate elements; if these plates have thickness t and width L, the mass of the solid is, in terms of density of the solid, $m_s = \rho_s t L^2$. The volume of a cell is $V_{cell} \propto L^3$, so the honeycomb density is given by the mass of the solid in a cell divided by its volume. So $\frac{\rho}{\rho_s} \propto \frac{t}{L}$.

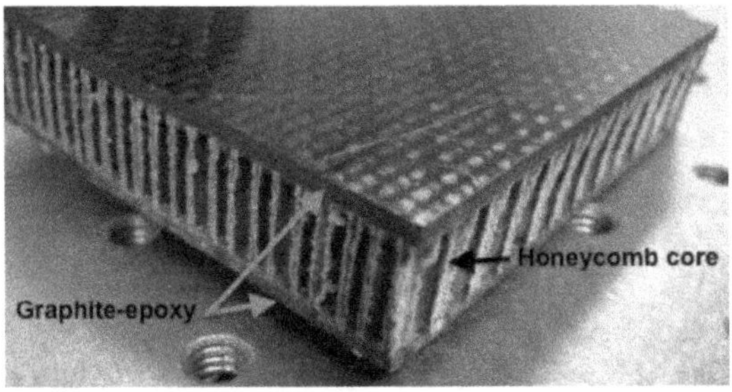

Figure 6.6: Sandwich panel structure with graphite epoxy face sheets and honeycomb core.

6.3.1 Honeycomb modulus

In-plane modulus: scaling

For honeycomb, subjected to in-plane deformation, as in Figure 6.5, the walls undergo plate bending, and the stiffness is sensitive to density, following a cubic dependence. The honeycomb has a thickness B so the plate elements have dimension B by L, t thick. The stress is $\sigma \propto FBL$ with F as load. The deflection u in terms of the Young's modulus of the solid is $u \propto FL^3/12E_sI$ with $I = bt^3/12$. The strain is $\epsilon \propto u/L$. The in-plane elastic modulus of the honeycomb is $E = \sigma/\epsilon \propto E_s\frac{t^3}{L^3}$. The constant of proportionality is obtained via a detailed analysis for the particular geometry.

In-plane modulus: exact

Force applied in-plane causes ribs to bend. Stress σ_1 in the x direction causes the ribs of length L to bend. The force

$$F = \sigma_1(H + Lsin\theta)B \tag{6.1}$$

gives rise to a bending moment

$$M = \frac{1}{2}FLsin\theta. \tag{6.2}$$

The resulting deflection u via beam theory, assuming the cell wall is slender, is

$$u = \frac{FL^3sin\theta}{12E_sI} \tag{6.3}$$

in which I is the area moment of inertia. For a cell wall of thickness t, $I = \frac{1}{12}Bt^3$. The component of deflection along the x axis is $u_x = usin\theta$ so the strain is $\epsilon_x = \frac{usin\theta}{Lcos\theta}$. Substituting,

$$\epsilon_x = \frac{\sigma_1(H + Lsin\theta)BL^2sin^2\theta}{12E_sIcos\theta}. \tag{6.4}$$

The Young's modulus for load in the x direction is

$$\frac{E_x}{E_s} = (\frac{t}{L})^3\frac{cos\theta}{(H + Lsin\theta)sin^2\theta}. \tag{6.5}$$

Similarly, load in the y direction also causes the ribs of length L to bend.

The strain in the y direction is

$$\epsilon_y = \frac{\sigma_2BL^4cos^3\theta}{12E_sI(H + Lsin\theta)}. \tag{6.6}$$

The Young's modulus for load in the y direction is

$$\frac{E_y}{E_s} = (\frac{t}{L})^3\frac{H + Lsin\theta}{cos^3\theta}. \tag{6.7}$$

For regular hexagons, $\theta = 30°$ and $H = L$, so

$$\frac{E_y}{E_s} = \frac{E_x}{E_s} = 2.3(\frac{t}{L})^3. \tag{6.8}$$

The behavior in-plane is two-dimensionally isotropic if the honeycomb is made of regular hexagonal cells.

Out-of-plane modulus

For honeycomb deformed out-of-plane, under tension or compression perpendicular to the honeycomb, the walls undergo axial deformation and the stiffness vs. density follows a Voigt model. The elastic modulus for force applied out-of-plane is

$$\frac{E}{E_s} = [\frac{\rho}{\rho_s}]. \tag{6.9}$$

Because the in-plane and out-of-plane moduli are so different, the honeycomb is highly anisotropic.

6.3.2 Honeycomb Poisson's ratio

In-plane Poisson's ratio

The in-plane Poisson's ratio is obtained from the ratio of the strains along and orthogonal to the direction of load.

$$\nu_{xy} = -\frac{\epsilon_y}{\epsilon_x} = \frac{cos^2\theta}{(\frac{H}{L} + sin\theta)sin\theta}. \tag{6.10}$$

$$\nu_{yx} = -\frac{\epsilon_x}{\epsilon_y} = \frac{(\frac{H}{L} + sin\theta)sin\theta}{cos^2\theta}. \tag{6.11}$$

For regular hexagons, $\theta = 30°$ and $H = L$; the Poisson's ratios are equal.

$$\nu_{xy} = \nu_{yx} = +1. \tag{6.12}$$

A value greater than $\frac{1}{2}$ is permissible because the material is anisotropic. If $\theta \to 0$, one Poisson's ratio tends to zero and the other Poisson's ratio becomes large. If $\theta < 0$, Poisson's ratio can be negative [13] in-plane.

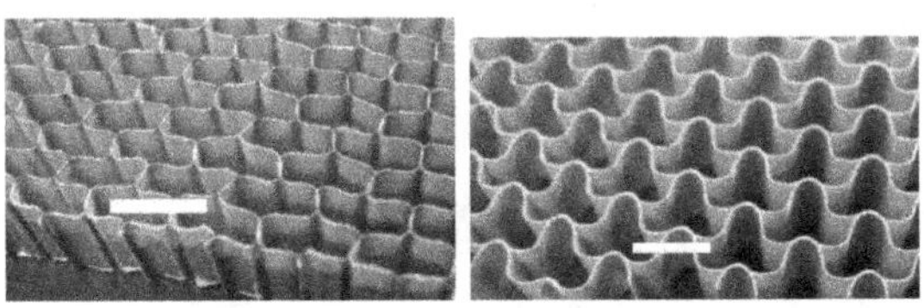

Figure 6.7: Honeycomb core for curved sandwich panels. Left, ox-core; right, flex-core. Scale bars: 10 mm.

Honeycomb with regular hexagonal cells is used as core for flat sandwich panels. The bending of such a honeycomb results in a saddle shape due to the Poisson effect. That is not helpful if one intends to design convex or cylindrical sandwich structures. The Poisson's ratio of the honeycomb can be changed by changing the shape of the cells (Figure 6.7), for example using elongated rather than regular hexagons [14]. These cells are so elongated that they appear rectangular. Such honeycomb is available commercially and is called ox-core. It is suitable for cylindrical sandwich structures. Cells shaped in the form of sombrero hats provide a positive Poisson's ratio but a value less than the maximal value +1. Such honeycomb can be deformed into a convex shape without buckling or crushing. Such honeycomb is available commercially and is called flex-core.

Out-of-plane Poisson's ratio

Suppose force is applied perpendicular to the plane of a honeycomb, so the stress is σ_{zz} with all other stresses zero. To show that the Poisson's ratio of the honeycomb for this direction is the same as that of the solid, ν_s, consider first a solid block of material under load in the z direction. Drill a hole in the z direction. There is no stress on the surface of the hole because the surface is perpendicular to z, so the Poisson effect is unchanged. The same is true if one drills an array of parallel holes. The honeycomb consists of a block with an array of hexagonal holes with parallel axes. Therefore the honeycomb Poisson's ratio for such deformation is the same as that of the solid from which it is made,

$$\nu_{zy} = \nu_{zx} = \nu_s. \tag{6.13}$$

6.3.3 Honeycomb strength

In-plane: elastic buckling

The cell walls of honeycomb behave as columns that can undergo buckling in compression. The strength is obtained from the critical Euler buckling load F_{crit} for a column of length L and cross-sectional moment of inertia I

$$F_{crit} = \frac{n^2\pi^2 E_s I}{L^2}. \tag{6.14}$$

n is related to the degree of constraint on the ends.

The load is $F = 2\sigma_2 LB cos\theta$. The buckling stress is obtained from the sum of the column loads divided by the area; the length is taken as H. For regular hexagons, $n = 0.69$.

$$\frac{\sigma^h_{el,z}}{E_s} = 0.22\left(\frac{t}{L}\right)^3. \tag{6.15}$$

Comparing this with Young's modulus, the collapse strain for load in-plane is 0.1.

Out-of-plane: elastic buckling

The strength for out-of-plane compression is most pertinent to the use of honeycomb in sandwich structures. As with the in-plane buckling strength, the column buckling Equation 6.14 is used.

The factor $1 - \nu_s^2$ comes from constraint associated with a compressed plate in contrast to a column. As for end constraint, an intermediate value of the constraint parameter is used.

$$\frac{\sigma_{el,z}^h}{E_s} = \frac{2}{1-\nu_s^2}\frac{\frac{L}{H}+2}{(\frac{H}{L}+sin\theta)cos\theta}\left(\frac{t}{L}\right)^3. \tag{6.16}$$

For regular hexagons, $\theta = 30°$ and $H = L$ and assuming for the solid $\nu_s = 0.3$, so

$$\frac{\sigma_{el,z}^h}{E_s} = 5.2\left(\frac{t}{L}\right)^3. \tag{6.17}$$

Out-of-plane: plastic buckling

The strength is obtained from the study of plastic hinge formation [1]. For honeycomb with regular hexagonal cells,

$$\frac{\sigma_{pl,z}^h}{\sigma_{y,s}} = 5.6\left(\frac{t}{L}\right)^{5/3}. \tag{6.18}$$

The difference in the power to which $\frac{t}{L}$ is raised implies elastic buckling will govern the strength if the cell walls are sufficiently slender.

Out-of-plane: yield

In a honeycomb if the net section stress in the plate elements along their long axis exceeds the yield stress $\sigma_{y,s}$ of the solid material comprising the cell walls, the walls will yield. The stress is just the force divided by the cross-sectional area of cell walls.

So

$$\frac{\sigma_{pl}^h}{\sigma_{y,s}} = \left[\frac{\rho}{\rho_s}\right]. \tag{6.19}$$

This limit will not be reached in compression unless the cell walls are very thick. Otherwise elastic or plastic buckling will occur first. To express this in terms of $\frac{t}{L}$ recall that for honeycomb with regular hexagonal cells $\frac{\rho}{\rho_s} = C_1\frac{t}{L}$ with $C_1 = 2/\sqrt{3}$.

6.3.4 Square cell honeycombs

Consider honeycomb with square cells (Figure 6.8). The relative density assuming thin plate elements, is $\frac{\rho}{\rho_s} = 2\frac{t}{L}$. To show this, observe that the cross-sectional area of a cell is L^2. The solid portion arises from the four cell walls of cross-section area tL. Each wall is shared by two adjacent cells, so $\frac{\rho}{\rho_s} = \frac{4}{2}\frac{tL}{L^2} = 2\frac{t}{L}$. The density of square honeycomb is in contrast to that of honeycomb of regular hexagons for which $\frac{\rho}{\rho_s} = \frac{2}{\sqrt{3}}\frac{t}{L}$.

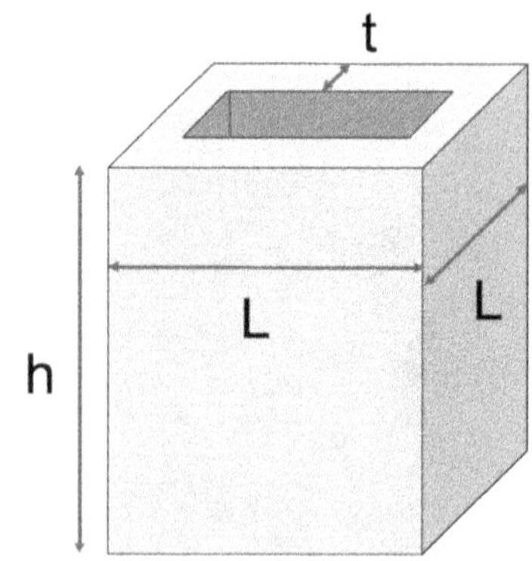

Figure 6.8: Square honeycomb cell.

The elastic modulus in the out-of-plane (z) direction is, by a Voigt construction, $\frac{E_3}{E_s} = [\frac{\rho}{\rho_s}]$. The elastic modulus in the in-plane (x and y) directions is $\frac{E_1}{E_s} = \frac{E_2}{E_s} = \frac{1}{2}[\frac{\rho}{\rho_s}] = \frac{t}{L}$. The factor $\frac{1}{2}$ is from the fact that half the plates are orthogonal to the load and do not contribute to the stiffness. This honeycomb is much stiffer in-plane than a hexagonal honeycomb for load along either principal axis in-plane; $\frac{t}{L} >> (\frac{t}{L})^3$. The square lattice in-plane is stretch-dominated while the regular hexagonal honeycomb is bend-dominated in all in-plane directions. The in-plane properties are governed by two-dimensional cubic symmetry which allows anisotropy in-plane. Indeed, the modulus in an oblique direction is much less than in a principal direction. Analysis similar to that for hexagonal honeycomb gives, for a 45° angle, $\frac{E_{45^\circ}}{E_s} = 2(\frac{t}{L})^3$.

Poisson's ratio for compression in the out-of-plane (z) direction is given by $\nu_{31} = \nu_{32} = \nu_s$. The reason is that compression of a solid block reveals a Poisson's ratio ν_s. Suppose a hole of any cross-section shape with surfaces parallel to the applied load is made in the solid. There is no stress component on these surfaces so the Poisson deformation is not affected. The same is true for an array of holes such as the voids in a honeycomb.

6.3.5 Hierarchical solids: honeycombs

Hierarchical solids contain structural elements which themselves have structure [52]. Designed honeycombs with structural hierarchy can exhibit enhanced properties, especially compressive strength. Large scale structural hierarchy is well known in 3D structures such as the Eiffel Tower, and the Garabit Viaduct designed by Eiffel; also in modern bridges and other structures. By using composite theory, one can design honeycomb, foam,

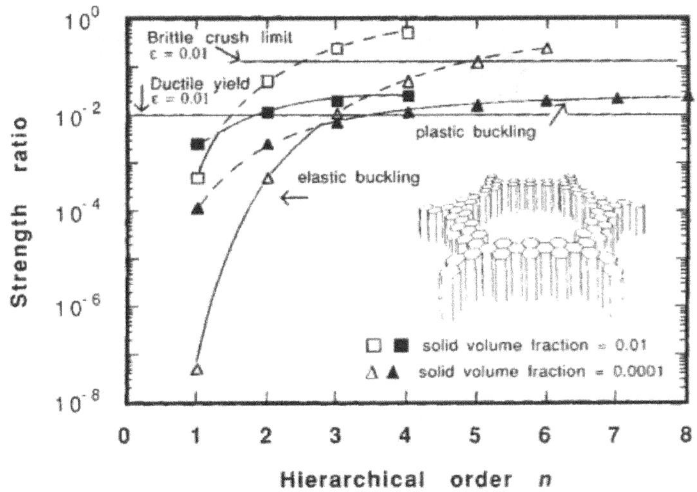

Figure 6.9: Strength enhancement of hierarchical honeycomb [52], with permission. Inset: second order honeycomb.

and lattice structures with enhanced properties. For example, hierarchical honeycombs and lattices can exhibit greatly enhanced (by a factor of 100 or more) compressive strength because buckling is suppressed as a result of the increased wall thickness [52]. Figure 6.9 shows a second-order honeycomb and predicted strength enhancement; such a structure was also made and studied experimentally.

6.3.6 Making honeycomb

Polymer-based honeycomb is made by preparing strips of the rib material. Stripes of glue are applied to each strip, then a stack of such strips is pressed together until the glue cures. This stack is called a HOBE block, an abbreviation of honeycomb before expansion. The block is then stretched or expanded, forming the honeycomb. Some cell walls are therefore of double thickness; this is taken into account in more sophisticated analysis.

Polymer-based honeycomb ribs are typically made of polymers called aramid; Nomex is related to nylon and Korek is related to Kevlar.

Metal-based honeycomb can be prepared by extrusion.

6.4 Foams

Foams are three-dimensional cellular solids. A foam in which the space within each cell communicates with the interior of adjacent cells is called

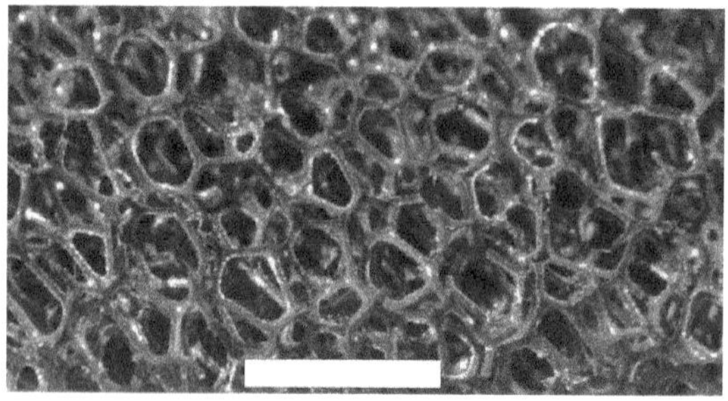

Figure 6.10: Open-cell polymer foam; scale bar 5 mm.

open-cell (Figure 6.10). The solid phase forms rib elements at the edges of the cells. A foam in which walls or membranes of solid phase separate adjacent cells is called closed-cell. Open-cell foams allow the free passage of fluids such as water or air through the structure.

6.4.1 Foam elastic modulus

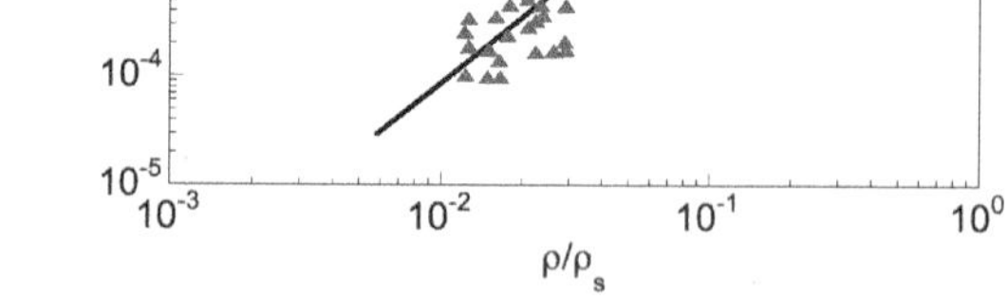

Figure 6.11: Young's modulus vs. density for foams, normalized to the modulus E_s of the solid comprising the ribs, adapted from [1]. Points are experimental; the solid line is based on analysis assuming bending of ribs.

To show that the stiffness of low density open-cell foam is proportional to the square of the relative density, consider the bending of the cell ribs. At small strains in the linear regime of behavior, open-cell foam deforms by the bending of the ribs or struts in the foam structure. Open-cell foam consists of rib elements; if these ribs have thickness t and length L, the mass of the solid is, in terms of density of the solid, $m_s = \rho_s t^2 L$. The volume of a cell is $V_{cell} \propto L^3$, so the foam density is given by the mass of the solid in a cell divided by its volume. So $\frac{\rho}{\rho_s} \propto \frac{t^2}{L^2}$.

The foam cells have a size proportional to rib length L; ribs are t thick. The stress is $\sigma \propto F/L^2$ with F as load. The deflection u in terms of the Young's modulus of the solid is via Equation 6.3, $u \propto FL^3/12E_sI$ with $I \propto t^4$. The strain is $\epsilon \propto u/L$. The elastic modulus of the foam is $E = \sigma/\epsilon \propto E_s\frac{t^4}{L^4}$. But as discussed in the prior paragraph, $\frac{\rho}{\rho_s} = \frac{t^2}{L^2}$. So the modulus is

$$\frac{E}{E_s} = C_f\left[\frac{\rho}{\rho_s}\right]^2, \tag{6.20}$$

in which ρ_s is the density, C_f is a constant of value near one, and E_s is the modulus of the solid material of which the foam is made. Rib bending gives rise to a quadratic dependence of foam modulus on density via elementary bending analysis. The analysis (solid line) is in agreement with experiments, Figure 6.11. The constant of proportionality is obtained via a detailed analysis for the particular geometry. For most isotropic foams the constant is close to 1. The determination of the constant C_f requires a detailed analysis of the foam cell geometry. For example, the foam cell may be modeled as the straight edges of a regular tetrakaidecahedron (polyhedron of 14 faces).

The quadratic dependence of modulus on density is easy to show via modeling the ribs using elementary beam theory. However, rigorous demonstration of the value of C_f requires a full three-dimensional analysis of the 14 side cell [15]. A repeating array of such cells is actually a lattice (§6.5.1), not a foam.

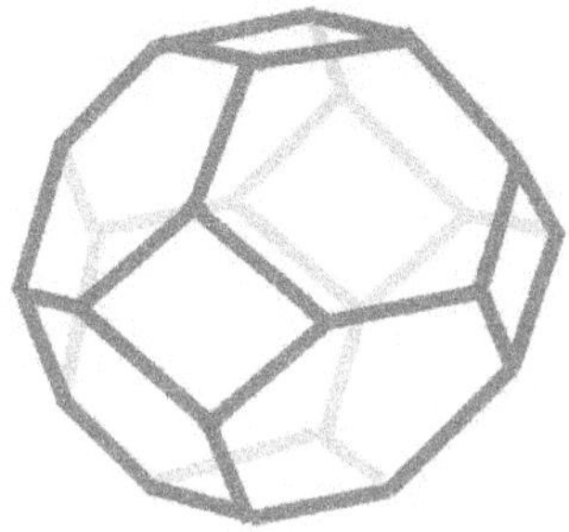

Figure 6.12: Ideal tetrakaidecahedron lattice cell.

The cell structure of foams is not periodic but it is not completely random. The cells of most synthetic low density polymer foams have a shape which can be modeled by the Kelvin minimum area tetrakaidecahedron.

This polyhedron, also called a truncated octahedron, depicted in Figure 6.12, is convex and has eight hexagonal faces and six square faces. These polyhedra were studied as space filling (or tiling of space, §6.2), by Lord Kelvin [16] in 1887. Foams considered at that time were of soap bubbles. Other polyhedral models were found to be unstable to the effects of surface tension. The ribs are straight in idealized models but they are curved in actual foams. The actual cell structure in a foam is not perfectly regular as shown in Figure 6.10.

Closed-cell low density foam usually has cell walls that are thin compared to the ribs as a result of the method of preparation. For such foam, the cell walls may be treated as membranes in the analysis of mechanical properties or ignored entirely. Compliant closed-cell low density foam has a contribution to stiffness from the compression of the air in the cells.

6.4.2 Poisson's ratio of foams

The Poisson's ratio of most foams in the linear regime is on the order 1/3 provided they are isotropic or nearly so. The Poisson effect occurs via reorientation of the cell ribs during deformation. Analysis of tetrakaidecahedron cells [15] predicts a large bulk modulus and a Poisson's ratio tending to 0.5. This is not confirmed by experiment in foams. The theory, though mathematically elegant, makes the simplifying assumptions that the foam ribs are straight and that the cells repeat ideally. Such a structure is actually a lattice (§6.5.1). Therefore the ribs in the model deform axially when the foam is subject to hydrostatic stress, hence a prediction of bulk modulus linear in the relative density. The ribs deform in bending if the foam is sheared or deformed axially, hence a prediction of G and E quadratic in the relative density. In reality, the foam ribs are curved, so they deform in bending in all modes of macroscopic deformation [17], so the bulk modulus does not greatly exceed the shear modulus, and the Poisson's ratio is close to 1/3, not 1/2.

6.4.3 Nonlinearity and strength of foams

A representative stress-strain curve for an open-cell elastomer foam (Figure 6.10) is shown in Figure 6.13. At a compressive strain above about 5%, the ribs in a flexible foam undergo elastic buckling, which gives rise to a plateau region in which the foam deforms progressively at near constant stress. At a sufficiently high compressive stress, the cell ribs come in contact and the foam densifies, which gives rise to a rapid increase in apparent stiffness. Under large deformation, foams behave nonlinearly, and the stress-strain diagram becomes distorted, as shown in Figure 6.13. If the foam is made of metal, the ribs may undergo plastic buckling; if the foam is made of ceramic,

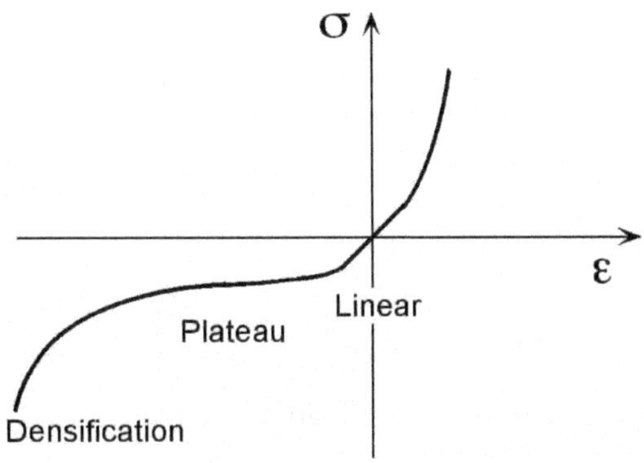

Figure 6.13: Representative stress-strain curve for an open-cell elastomer foam.

the ribs may fail by crushing. In such materials the overall shape of the linear, plateau and densification regions is similar to those for elastomer. Poisson's ratio for large strain in compression becomes small due to the buckling of ribs. For large strain in tension, Poisson's ratio can exceed 1/2 due to rib realignment and induced anisotropy.

If a wider linear range of deformation in polymer foam is desired, as in various pads, foam may be modified by the Omelon process in which the foam is permanently compressed in one direction. This process buckles the ribs and freezes them in the buckled configuration. The change in slope of the load-deformation curve that would occur at about 5% strain is thereby eliminated.

To determine the compressive strength of foam based on the elastic buckling of ribs, use Equation 6.14 and the critical load F_{crit} for column buckling. The strength for buckling is

$$\sigma_{el} \propto \frac{F_{crit}}{L^2} \propto \frac{E_s I}{t^4}. \tag{6.21}$$

But $I \propto t^4$; also $\frac{\rho}{\rho_s} = \frac{t^2}{L^2}$ So

$$\frac{\sigma^f_{el,z}}{E_s} = C_f \left(\frac{\rho}{\rho_s}\right)^2. \tag{6.22}$$

The proportionality constant can be obtained from experiment as $C_f = 0.05$. A similar value is obtained from the analysis of rectangular frameworks of beams and columns.

6.4.4 Toughness of foams

The toughness is calculated by considering a crack in the foam under tension, and analyzing the bending of the cell ribs. The toughness is given by the following [1], with σ_s^{ult} as the strength of the solid and ℓ_{cell} as the cell size. Observe that toughness depends on structure size. This is in contrast with the theory of elasticity, which has no length scale within the theory.

$$K_{1c} = C_t \sigma_s^{ult} \sqrt{\pi \ell_{cell}} \left[\frac{\rho}{\rho_s} \right]^{3/2} \tag{6.23}$$

with $C_t = 0.65$. The theoretical dependence of toughness on density (solid line) is supported by experiments (points) by various authors as shown in Figure 6.14.

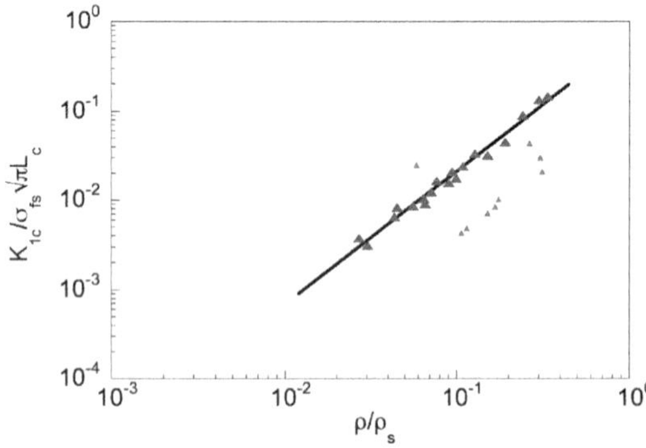

Figure 6.14: Toughness of foam vs. relative density, adapted from [1].

6.4.5 Dense foams and syntactic foams

High-density rigid polyurethane closed-cell foam, called last-a-foam, is commercially available [18]. It is used for composite panel cores, flotation devices, and other uses. A density of 360 kg/m^3 corresponds to 20 pounds per cubic foot; densities of 10 to 30 pounds per cubic foot are available. The cells are typically smaller than 0.15 mm in size. Foams which have cells too small to be evident to the unaided eye are called microcellular. For such dense foams, the structure (Figure 6.15) consists of voids as bubbles rather than a network of ribs. Rib-based analysis does not apply to such foams.

Syntactic foams consist of hollow spheres, typically glass, embedded in a polymer matrix [19] [20]. The inclusion-matrix morphology resembles that of a particle reinforced composite. The hollow spheres, by their symmetry, resist hydrostatic compression. The empty space within the spheres

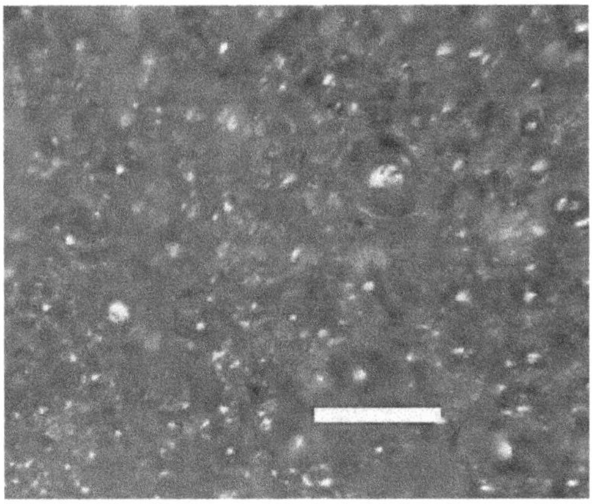

Figure 6.15: Dense polyurethane foam, optical micrograph. Scale mark, 0.25 mm.

reduces the density. Representative properties are density 42 lb/ft^3 (0.67 g/cm^3), compressive modulus 2.9 GPa, compressive strength 110 MPa, hydrostatic crush strength 20,000 psi (1.4 kbar or 138 MPa) [19]. They are used in applications in which underwater buoyancy is required; the high crush strength allows use at great depth. They are also used as structural core in sandwich structures for aerospace applications, and in some sporting goods.

6.4.6 Making foams

Polymer foams are made by adding a foaming agent, also called a blowing agent, to the liquid monomer mix. The foaming agent reacts to evolve gas which forms the void space in the foam. Water is often used as the foaming agent; water combined with a monomer such as di-isocyanate evolves carbon dioxide gas which expands the foam. The liquid monomer is also mixed with chemicals that initiate the polymerization reaction to create long chain molecules. Chemicals are forced into a mixing head and are poured into a conveyor system in which both bottom and lateral surfaces move. Blocks up to 54 inches wide and of arbitrary length are made in a continuous process. The fraction of open cells is controlled by catalyst, surfactant and processing pressure. If a foam with all open cells is desired, remaining closed-cell walls may be broken with a pulse of air or by mechanical pressure. The foaming process is exothermic; temperatures can exceed 300°F. Most commercial low density foams are 30 to 80 pores per inch (0.85 mm to 0.32 mm cells); foams with 10 to 100 pores per inch (2.5 mm to 0.25 mm cells) are available. As for density, the typical range for compliant polymer foam is typically 1 to 4 pounds per cubic foot (0.016 g/cc to 0.065 g/cc).

6.5 Lattices

Lattices are periodic arrays of structural elements. If these elements are atoms, the lattice is a crystal. In the present context, the structural elements are rib, plate or curved surface elements that can be viewed as an elastic or a viscoelastic continuum. If the ribs are straight and slender, analysis is simplified as is the case with foams. Honeycombs (§6.3) are two-dimensional lattices. In the following, three-dimensional lattices are considered.

Three-dimensional elastic lattices of slender straight ribs can be made on a macroscopic scale (in building structures) or on a fine scale via additive manufacturing methods such as 3D printing and related methods. Lattices with cubic structural symmetry are simpler than those with less symmetry; they are reviewed in the following. It is possible to design three-dimensional lattices that are are stiffer than foams made of the same rib material if the lattice ribs are organized in a triangulated structure such that the ribs deform axially rather than in bending. Such structures are called stretch-dominated. Lattices containing plate or surface elements can be stiffer and stronger than lattices containing ribs.

6.5.1 Truss lattices: ribs

Lattices made of rib elements are called rib lattices or truss lattices. Lattices can be envisaged based on the tessellation of space by identical geometrical figures as discussed in §6.2 or by a combination of geometrical figures. Consider a simple 3D cubic lattice comprised of straight slender ribs (Figure 6.16, left). Lattices containing rib elements are called truss lattices. The Young's modulus of the simple cubic lattice in a principal direction is governed by a Voigt construction. One third of the ribs are aligned with the force, so $\frac{E}{E_s} = \frac{1}{3}[\frac{\rho}{\rho_s}]$. Simple cubic lattices are anisotropic. To obtain isotropic elastic response, oblique elements are used. For example, one may combine ribs of a simple cubic truss lattice with ribs of a body-centered truss lattice to achieve elastic isotropy. For an elastically isotropic truss lattice [21] [22], $\frac{E}{E_s} = \frac{1}{6}[\frac{\rho}{\rho_s}]$.

A lattice of tetrakaidecahedron (14 face polyhedron) cells, Figure 6.12, is stiff under hydrostatic stress but is compliant under uniaxial or shear stress for which the rib deformation is bend-dominated. Such cells have been studied to model foam [15]; the analysis is more pertinent to lattices which have straight ribs and repeating structure. The bulk modulus is

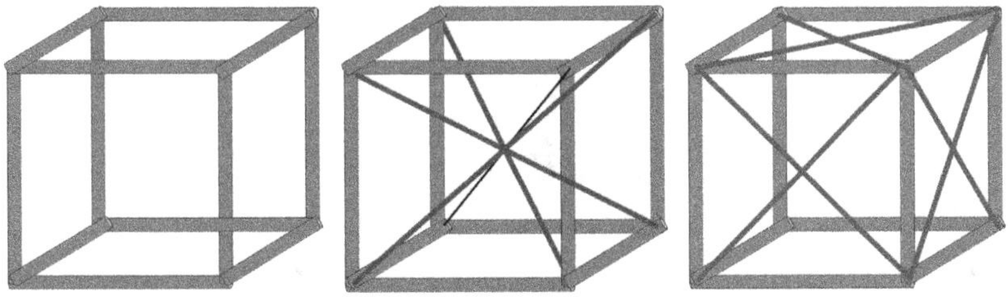

Figure 6.16: Cubic lattice cells. Left, simple cubic truss lattice; center, body-centered cubic truss lattice; right, face-centered cubic truss lattice.

governed by stretching of the ribs [15]. The bulk modulus K normalized to Young's modulus E_s of the solid ribs is $\frac{K}{E_s} = \frac{1}{9}[\frac{\rho}{\rho_s}]$. The Young's modulus E is governed by the bending of ribs. For ribs of a Plateau cross-section associated with surface tension in foam, and for small density,

$$\frac{E}{E_s} = 1.009\left[\frac{\rho}{\rho_s}\right]^2. \tag{6.24}$$

Lattices of tetrakaidecahedron cells have been considered to approximate the structure of open-cell foams. However the lattice Poisson's ratio approaches $\frac{1}{2}$ compared with $\frac{1}{3}$ for most foams.

Rib bending is sensitive to the cross-section shape. If the ribs have a circular cross-section and a Poisson's ratio of zero then the tetrakaidecahedron lattice is elastically isotropic [23]. If the ribs have a Plateau cross-section [15] the elastic properties are nearly isotropic. The Plateau section has inwardly curved surfaces. These result from surface tension during foam formation.

Representative properties of several lattices with cubic symmetry are given in Table 6.1. The rib radius is r. The length of a rib in the unit cell is L; this is not necessarily the length of the rib in the structure. For these lattices the bulk modulus [23] normalized to Young's modulus E_s of the solid ribs is $\frac{K}{E_s} = \frac{1}{9}[\frac{\rho}{\rho_s}]$.

C moduli [24] for a face-centered cubic (FCC) truss lattice and for a simple cubic plate lattice (§6.5.4) are given in Table 6.2. Lattice C moduli were also presented in [25]. Referring to Table 6.2, the modulus C_{1111} is about $1/6$ E_s for the FCC truss lattice and about $2/3$ E_s for the simple cubic plate lattice. The anisotropy ratio is 1.21 for the FCC truss lattice and is 2.4 for the plate lattice if Poisson's ratio is 0.3.

The octet truss, originally after Buckminster Fuller [10], is based on a tessellation containing regular octahedra and tetrahedra as shown in Figure 6.17. This structure can be obtained from the face-centered cubic (FCC)

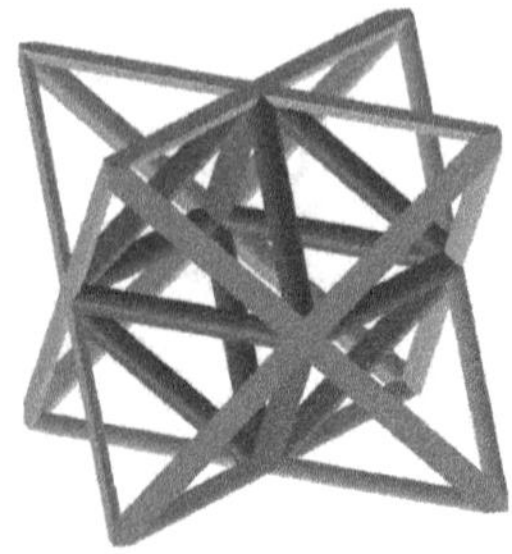

Figure 6.17: Ideal octet lattice cell [26].

Table 6.1: Truss lattice material properties: bulk modulus K and shear modulus G [23]. L refers to length of rib in unit cell; $\bar{L}$ is average length. Ribs are assumed to be slender.

Structure	$[\frac{\rho}{\rho_s}]$	K	G_1	G_2
simple cubic	$2.36(\frac{r}{L})^2$	$E_s\frac{1}{9}\frac{\rho}{\rho_s}$	$E_s\frac{3}{2\pi}\frac{1}{9}(\frac{\rho}{\rho_s})^2$	$E_s\frac{3}{2}\frac{1}{9}\frac{\rho}{\rho_s}$
octet	$6.66(\frac{r}{L})^2$	$E_s\frac{1}{9}\frac{\rho}{\rho_s}$	$E_s\frac{3}{4}\frac{1}{9}\frac{\rho}{\rho_s}$	$E_s\frac{3}{8}\frac{1}{9}\frac{\rho}{\rho_s}$
tetrakaidecahedral	$8.66(\frac{r}{\bar{L}})^2$	$E_s\frac{1}{9}\frac{\rho}{\rho_s}$	$E_s\frac{4\sqrt{2}}{\pi}\frac{1}{9}(\frac{\rho}{\rho_s})^2$	$E_s\frac{4\sqrt{2}}{\pi}\frac{1}{9}(\frac{\rho}{\rho_s})^2$

lattice by connecting the nearest-neighbor lattice sites. The octet lattice is stretch-dominated under all stress states, so it is stiff for all stress states [25]. The octet lattice was also used for zero thermal expansion designs [27]. For the octet structure, properties are given [28] in the compliance formulation as $J_{11}^{-1} = E = \frac{1}{9}\frac{\rho}{\rho_s}E_s$; $J_{44}^{-1} = G = \frac{1}{12}\frac{\rho}{\rho_s}E_s$; $J_{12}^{-1} = \frac{1}{3}\frac{\rho}{\rho_s}E_s$. The criterion for isotropy is $J_{44} = 2(J_{11} + J_{12})$. As with other anisotropic solids, the rotation of coordinate axes with respect to the principal axes will produce nonzero stretch shear coupling terms. The octet truss lattice has cubic symmetry. In principal directions, $\frac{E}{E_s} = \frac{1}{9}[\frac{\rho}{\rho_s}]$. This material is elastically anisotropic. The rotation of coordinates gives moduli $\frac{E}{E_s} = \frac{1}{6}[\frac{\rho}{\rho_s}], \frac{1}{6}[\frac{\rho}{\rho_s}], \frac{1}{5}[\frac{\rho}{\rho_s}]$ in different directions. These are larger than in

Table 6.2: Lattice material properties for truss FCC and plate cubic lattices [24] of low density.

Structure	C_{1111}	C_{1122}	C_{1212}
Truss FCC	$\frac{1+\sqrt{2}}{3(1+2\sqrt{2})}(\frac{\rho}{\rho_s})E_s$	$\frac{1}{3(1+2\sqrt{2})}(\frac{\rho}{\rho_s})E_s$	$\frac{1}{3(1+2\sqrt{2})}(\frac{\rho}{\rho_s})E_s$
Plate cubic	$\frac{2}{3(1-\nu_s^2)}(\frac{\rho}{\rho_s})E_s$	$\frac{1}{3(1-\nu_s^2)}(\frac{\rho}{\rho_s})E_s$	$\frac{1}{3}(\frac{\rho}{\rho_s})G_s$

principal directions, at the expense of stretch-shear coupling terms and larger compliance in shear.

For *cubic* symmetry, $K = (C_{11} + 2C_{12})/3$, $G_1 = C_{44}$ (the principal shear modulus) and $G_2 = (C_{11} - C_{12})/2$; there are three independent elastic constants. Recall that for an *isotropic* material, $\frac{1}{2}(C_{11} - C_{12}) = C_{44}$ so there are two independent elastic constants and the shear moduli are equal, $G_2 = G_1$. The inequality of the shear moduli is a measure of the cubic anisotropy. The anisotropy may also be quantified by the Zener ratio [29], $Z = 2\frac{C_{44}}{C_{11}-C_{12}}$, which applies for cubic materials; isotropy entails $Z = 1$. The Zener ratio may be written $Z = \frac{2G(1+\nu)}{E}$.

Several truss lattices obtained from Archimedean polyhedra were studied [30] in the context of stiffness and strength. Lattices based on the cubo-octahedron and cubic lattices of FCC and BCC structure were found to be stretch-dominated and to have stiffness similar to the octet. Lattices based on the the truncated cube, the truncated octahedron, the great rhombicuboctahedron, and the small rhombicuboctahedron were found to be bend-dominated, so considerably more compliant.

Lattices have been made with snap-through characteristics for light weight shock absorbers [31] [32]. Such lattices follow the paradigm articulated in §2.5.2 in which the controlled instability of a nonlinear system in a negative stiffness region can give rise to large hysteresis to dissipate mechanical energy.

6.5.2 Continuous rib lattices

Figure 6.18: Polymer lattice provided by [33]. Scale bar: 10 mm.

An early polymer lattice made by additive manufacturing (Figure 6.18), [33], was made by an optical additive manufacturing method that generates straight ribs with no corners or changes in direction. The modulus ratio [33] is

$$\frac{E}{E_s} = sin^4\theta\left[\frac{\rho}{\rho_s}\right], \tag{6.25}$$

in which the constant of proportionality is due to ribs at an angle θ from the direction of load. The cell size in such lattices is typically several millimeters.

The density depends on the aspect ratio of radius to length via $\frac{\rho}{\rho_s} = 2\pi(r/L)^2[sin\theta cos^2\theta]^{-1}$. The compressive strength of such lattices is governed by the buckling of the ribs, as is the case for foams [33]; k depends on the end condition of the ribs.

$$\frac{\sigma_{el}}{E_s} = \left[\frac{\pi}{k}\right]^2 \left[\frac{r}{L}\right]^2. \tag{6.26}$$

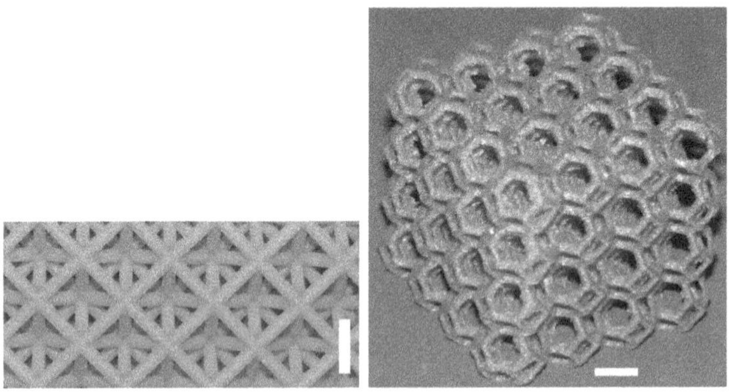

Figure 6.19: Polymer lattices. (a) Octet structure lattice, 3D printed [34]; scale bar 1 cm. (b) Tetrakaidecahedron structure lattice, 3D printed [35]; scale bar 1 cm.

The lattice shown in Figure 6.19 (a) consists of a cubic array of *octet truss* unit cells (Figure 6.17), providing stretch-dominated response and a high ratio of stiffness to density. Such lattices have been made with a cell size smaller than 0.3 mm [36]. The lattice in Figure 6.19 (b) contains cells of tetrakaidecahedron shape; although the cells are regular, it is bend-dominated like a foam. The density is 0.1 g/cm^3. Such lattices have also been made with a cell size smaller than 0.3 mm [36].

6.5.3 Lattice property bounds

The classic bounds for moduli of two-phase composites [37] can be used for cellular solids including lattices. For example the bulk modulus lower bound discussed in §2, becomes an upper bound by interchanging subscripts 1 and 2:

$$K_U = K_1 + \frac{V_2(K_2 - K_1)(3K_1 + 4G_1)}{(3K_1 + 4G_1) + 3V_1(K_2 - K_1)}. \tag{6.27}$$

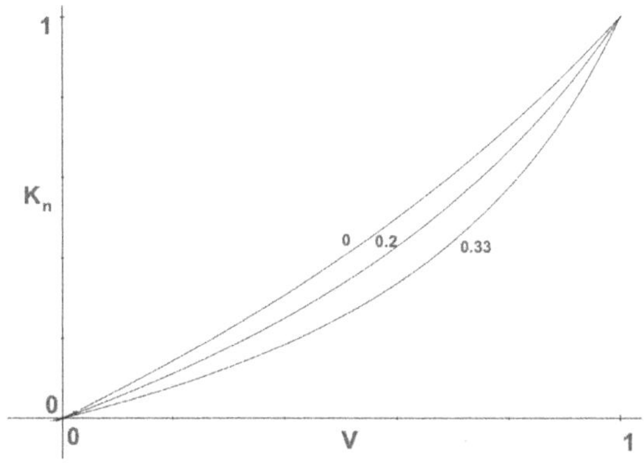

Figure 6.20: Upper bound on bulk modulus solid of volume fraction V and void space. Normalized bulk modulus $K_n = \frac{K_U}{K_1}$. Curves are for Poisson's ratios 0, 0.2, 0.33 of the solid material.

The lower bound for cellular solids is zero because if the stiff phase is surrounded by void space, there is no physical connection, hence no stiffness. That case corresponds to $K_1 = 0$. Now let $K_1 = 0$ in the upper bound and relate the shear modulus G to the bulk modulus K via Poisson's ratio ν:

$$G = K\frac{3}{2}\frac{(1-2\nu)}{1+\nu} \tag{6.28}$$

Figure 6.20 shows the bulk modulus upper bound normalized to the solid bulk modulus for several Poisson's ratios ν_s of the solid. The initial portions of the curves reveal the modulus for small V. One may also normalize to V to show the dependence on volume fraction for small concentration via the intercept with the ordinate. For example if $\nu_s = 0.2$, $K = 0.5VK_s$ for small V.

The Hashin-Shtrikman upper bounds for a cellular solid may be written [24] in terms of Young's modulus E_s and Poisson's ratio ν_s of the solid phase, provided the relative density ρ/ρ_s is small. The bound on the bulk modulus K is

$$K = \frac{2}{9(1-\nu_s)}\left(\frac{\rho}{\rho_s}\right)E_s. \tag{6.29}$$

The bound on the shear modulus G is

$$G = \frac{(7-5\nu)}{30(1-\nu_s^2)}\left(\frac{\rho}{\rho_s}\right)E_s. \tag{6.30}$$

The lower bounds are zero because the solid constituent may be disconnected in particulate form.

Comparing the bounds with truss lattices of a small volume fraction of the solid, and recalling that for a cellular solid, $V = \frac{\rho}{\rho_s}$, the truss lattices do not approach the upper bound. So even though truss lattices are stiffer than foams of the same density and solid constituent properties, they are not as stiff as theory allows. The following lattices, made of plate and surface elements, provide greater stiffness.

6.5.4 Plate lattices

A simple cubic lattice comprised of straight slender plate elements (Figure 6.21) is stiffer than a lattice of rib elements (of the same relative density) because more material is aligned with stress in principal directions. For this plate lattice, $\frac{E}{E_s} = \frac{2}{3}[\frac{\rho}{\rho_s}]$, again via a Voigt construction. This lattice has cubic structure and is elastically anisotropic. It is possible to design an isotropic plate lattice, consisting of superposed simple cubic and body-centered lattices [38]; at low density, $\frac{E}{E_s} = \frac{1}{2}[\frac{\rho}{\rho_s}]$. Such plate lattices at low density approach the theoretical upper bound on stiffness [37]. It has long been known that one can attain the upper bound on stiffness via hierarchical coated sphere or laminate morphology (§2.3.5); to obtain a cellular material, one constituent of the two-phase composite becomes empty space.

Isotropic lattices are not fully equivalent to a homogeneous material of the same modulus and density. At sufficiently high frequency, rib or plate elements can be set into vibration. Such vibration affects wave propagation and can lead to the cut-off of waves above a critical frequency. Also, lattices can differ from homogeneous solids when there are strain gradients such as those that occur in bending, torsion (§8.3) and in stress concentrations (§8.2).

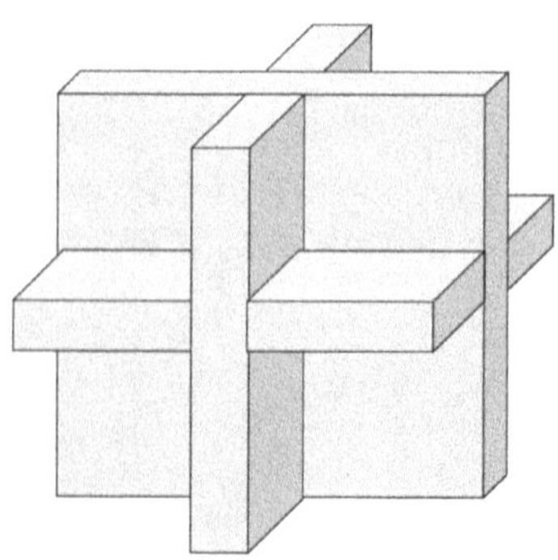

Figure 6.21: Cubic plate lattice cell: simple cubic plate lattice.

6.5.5 Surface lattices

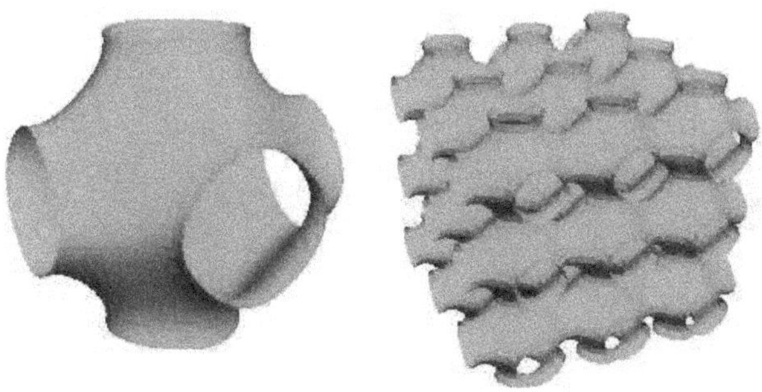

Figure 6.22: P lattice. Left, cell detail; right, lattice.

Lattices with shell-like members have long been known [43] [44]; they were claimed to deliver maximal strength per weight because stress is diffused via membrane action. For example, the P lattice shown in Figure 6.22 is describable [46] by the surface

$$cos 2\pi x + cos 2\pi y + cos 2\pi z = 0. \tag{6.31}$$

In a physical embodiment, a shell lattice is given a nonzero thickness that is comprised of solid material. Shell lattices have been more recently shown to perform better than truss lattices [39], even those that are stretch-dominated. The bulk modulus K of such lattices approaches the Hashin-Strikman upper bound. The bounds, expressed in terms of K and G can be related to Poisson's ratio via $K = \frac{2G(1+\nu)}{3(1-2\nu)}$. For small volume fraction and for $\nu = 0.3$, the bound is $\frac{K}{K_s} \approx 0.4[\frac{\rho}{\rho_s}]$. Recall for comparison that $E = 3K(1-2\nu)$ so for a solid Poisson's ratio of 0.2, $E = 1.8K$; for $\nu = 1/3$, $E = K$. Shell lattice Young's and shear moduli do not approach the upper bound except at high density; nevertheless they exceed the moduli of truss lattices. Both shell and truss lattices can be made elastically isotropic; shell lattices are stiffer. The strength of shell lattices exceeds the strength of the octet truss [39].

Gyroid lattices [40] [41], are shell lattices based on minimal surfaces [42]. The gyroid surface subdivides three-dimensional space into two parts called labyrinths [45]. Each labyrinth is connected but not simply connected. The gyroid surface is describable [46] by

$$sin 2\pi y cos 2\pi z + sin 2\pi z cos 2\pi x + sin 2\pi x cos 2\pi y = a \tag{6.32}$$

with a as a constant.

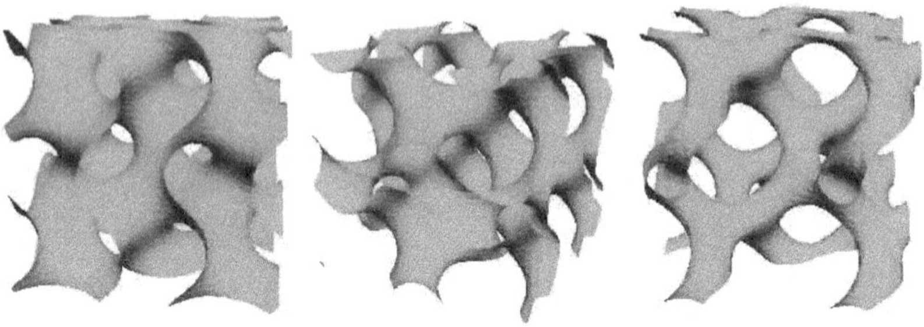

Figure 6.23: Gyroid lattices. Left to right, gyroid surfaces with a = 0, 0.6, 1.

The labyrinths associated with the gyroid surface (for $a = 0$) are enantiomorphic: they are mirror images of each other. The labyrinth shape is chiral for the gyroid surface. One may choose a value of a (Figure 6.23) and fill one of the regions with solid material, or one may embody the gyroid as a shell lattice with nonzero thickness.

Gyroid lattices have better mechanical properties than truss lattices with ribs. The surfaces can be given a nonzero thickness to allow control of volume fraction of solid [47]. Gyroid lattices have been developed for tissue growth scaffolds [48].

A comparison was made between a simple cubic lattice and a gyroid lattice. The Young's modulus of a simple cubic lattice along a principal direction is 324 MPa for structures with a porosity of 64%. A gyroid lattice with $a = 0.60$ and one spatial region filled with solid was made. The gyroid lattice was less stiff than a simple cubic lattice of similar geometry, 169 MPa at 69% porosity [48]. The simple cubic lattice is highly anisotropic in that it is compliant for stretching in oblique directions. The degree of elastic anisotropy of the gyroid lattice is not known. Gyroid lattices have also been studied for simultaneous transport of heat and electricity [49].

The gyroid lattice at low relative density has been idealized by a framework of slender ribs of circular cross-section, diameter d and length L [50]. The relative density is $\frac{\rho}{\rho_s} = \frac{3\pi}{16\sqrt{2}}(\frac{d}{L})^2$. The symmetry is cubic. The principal elastic moduli are Young's modulus $\frac{E}{E_s} = 0.426[\frac{\rho}{\rho_s}]^2$, shear modulus $\frac{G}{E_s} = 0.329[\frac{\rho}{\rho_s}]^2$, and bulk modulus $\frac{K}{E_s} = \frac{1}{9}[\frac{\rho}{\rho_s}]$. Axial and shear deformation of the gyroid rib lattice are governed by rib bending hence the quadratic dependence on density. Bulk deformation is governed by rib extension hence the linear dependence on density. Observe that the bulk modulus is the same as that predicted for octet, cubic and tetrakaidecahedron lattices as presented in Table 6.1. The strength of the gyroid rib lattice for hydrostatic stress is, for the elastic buckling of ribs, $\frac{\sigma_e^l}{E_s} = \frac{1}{13}[\frac{\rho}{\rho_s}]^2$; for the plastic buckling of ribs of strength $\sigma_{y,s}$ it is $\frac{\sigma_{pl}^f}{\sigma_{y,s}} = 0.44[\frac{\rho}{\rho_s}]^{3/2}$.

Table 6.3: Lattice material properties: Young's modulus E for lattices with slender ribs and plates.

Structure	E
cubic truss, principal	$E_s \frac{1}{3} \frac{\rho}{\rho_s}$
cubic plate, principal	$E_s \frac{2}{3} \frac{\rho}{\rho_s}$
octet truss principal	$E_s \frac{1}{9} \frac{\rho}{\rho_s}$
octet truss max	$E_s \frac{1}{5} \frac{\rho}{\rho_s}$
tetrakaidecahedral	$1.01\ E_s (\frac{\rho}{\rho_s})^2$
isotropic truss	$E_s \frac{1}{6} \frac{\rho}{\rho_s}$
isotropic plate	$E_s \frac{1}{2} \frac{\rho}{\rho_s}$

Young's moduli of several lattices are compared in Table 6.3. The octet lattice of ribs has cubic symmetry [28]. The lattice with tetrakaidecahedral cells also has cubic symmetry [15]. For axial deformation, the ribs of this lattice deform mostly by bending. As discussed above, the cross-section of the rib shape influences the lattice modulus. The value in the table assumes the ribs have a Plateau cross-section typical of open-cell foams; this shape has inwardly curved surfaces as a result of surface tension during foam formation. If instead the ribs are triangular in cross-section, the modulus is reduced to about 73% of the value for a Plateau section. The elastic behavior is nearly isotropic, with about a 4% variation of modulus with direction. Lattices based on plate elements are stiffer and stronger than lattices based on rib elements.

6.5.6 Hierarchical lattices

Hierarchical solids contain structural elements which themselves have structure [52]. They can be honeycomb (2D lattices) (§6.3.5) or 3D foams and lattices. As with 2D lattices, 3D lattices exhibit higher strength with structural hierarchy provided the structure is organized appropriately, as shown in Figure 6.24.

A hierarchical structure was used to attain the Hashin-Shtrikman bounds (§2.3.5) in two-phase composites. Structural hierarchy was used to design a laminate (Figure 2.9) [51] with a negative Poisson's ratio. Additive manufacturing methods (§6.5.7) such as 3D printing have enabled complex structures to be made without manual assembly.

Resilient 3D hierarchical structures have been made (Figure 6.25) by laser writing in photoresist [53] and studied experimentally. Tubular ribs

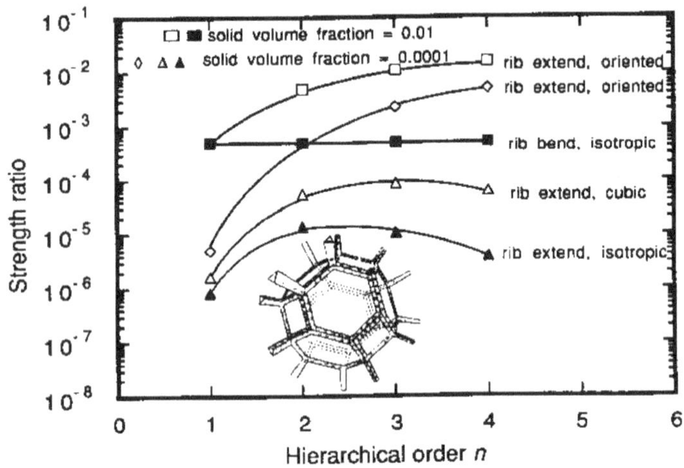

Figure 6.24: Strength enhancement of hierarchical 3D lattice [52], with permission. Inset: second order tetrakaidecahedral cell.

provide a beneficial level of structural hierarchy [54], Figure 6.26. The change from solid to hollow ribs effectively increases the hierarchical order by one. Such hierarchical structures may be regarded as approaching a fractal as hierarchical order increases.

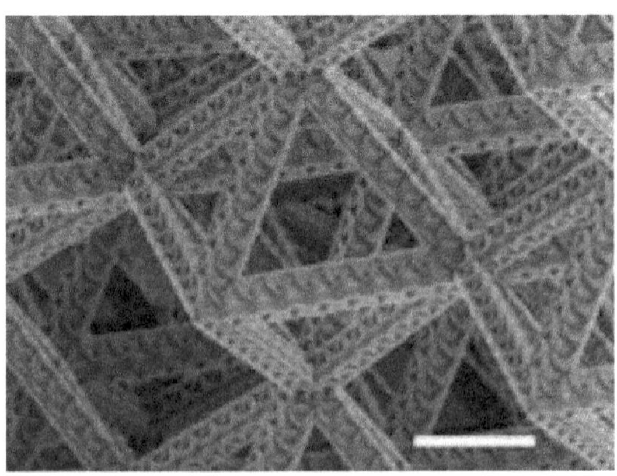

Figure 6.25: Octahedral hierarchical lattice [53] with permission. Scale bar, 10 μm.

Lattices of ultra-low density (less than 10 milligrams per cubic centimeter) were made with thin wall metal tubes [55]. Such lattices may be of use

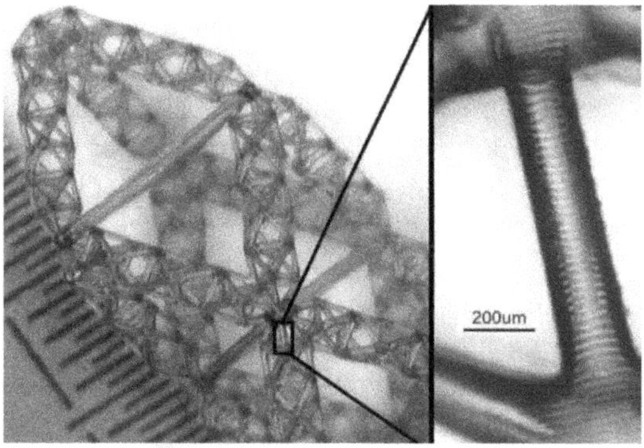

Figure 6.26: Tubular hierarchical lattice [54] with permission. Scale bar, 200 μm.

in thermal insulation; battery electrodes; catalyst supports; and damping of acoustic waves, vibration, or impacts. To make the lattices, a template was formed by photopolymer waveguide prototyping. The template was coated with nickel by electroless plating. The polymer template was etched away leaving a lattice of thin wall nickel tubes as shown in Figure 6.25. The lattices exhibited complete recovery after compression exceeding 50% strain, and energy absorption similar to elastomers. The properties are attributed to structural hierarchy at multiple length scales.

Figure 6.27: Ultra-low density lattice with tubular metal ribs, of type studied in [55], after [56], with permission. Scale bar, 5 mm.

6.5.7 Making lattices

There are several methods for making lattices; this is a subject of continuing development. Most such methods are described as 3D printing, additive manufacturing, or rapid prototyping. In such methods, a conceptual 3D model is drawn using computer software and exported to the printer.

The fused deposition modeling (FDM) method makes use of heat to liquefy a thermoplastic polymer. The thermoplastic is then extruded in a fine bead along a path governed by the control electronics. Support structures are often necessary; these can be broken off or dissolved after the lattice has been fabricated. Stereolithography (SLA) makes use of an ultraviolet laser controlled by the electronics to build a 3D structure layer by layer by polymerizing liquid polymer in a tank.

In a variant laser method, polymer lattices were made by using a light-sensitive monomer liquid. They are made by using controlled beams of light to polymerize a precursor liquid containing monomer [33]. The lattice structure forms according to the pattern of the light beams. This method is therefore limited because light beams cannot turn corners. The remaining monomer is drained off, leaving the lattice.

Lattices can also be made in polymers, metals or ceramics by additive methods such as projection stereo-lithography [36]. In this method, a polymer is used similarly to the above. Each image is projected through a reducing lens on the surface of photosensitive resin. The exposed layer and its substrate are lowered, and a thin film of liquid is applied over the cured layer. The image projection is then repeated to form additional layers. If desired, then metal can be plated on the surface by electroless plating. The polymer is then removed by thermal decomposition, leaving behind a lattice of thin metal tubes. Ceramics can be made by atomic layer gas phase deposition on the polymer substrate; the ceramic overlay is 40 to 200 nm thick. The field of additive manufacturing is rapidly evolving.

6.6 Poisson's ratio tuning

6.6.1 Poisson's ratio in anisotropic materials

Prior to the development of negative Poisson's ratio foams, Poisson's ratio in materials had been viewed as an intrinsic property, not subject to modification. The earliest known report of negative Poisson's ratio is in the book on elasticity by Love [57]: $\nu = -1/7$ was suggested in some directions in anisotropic single crystal iron pyrites. Subsequent experiments [58] revealed a positive Poisson's ratio in this material. Single crystals of arsenic, antimony, and bismuth [59] exhibit a negative Poisson's ratio in some directions. These crystals are highly anisotropic. Arsenic has the greatest degree of anisotropy of these; it exhibits a factor of 11.3 variation in Young's modulus with direction. Arsenic crystals exhibit a Poisson's ratio that exceeds 0.5 or is less than −1, depending on the direction of stress. The anisotropy arises from a layered structure with compliant bonding between layers in the crystals. A negative Poisson's ratio can also occur in other highly anisotropic materials such as honeycomb with shaped cells [13]; they were thought to be unknown in 3D isotropic materials.

6.6.2 Poisson's tuning ratio in foams: negative or extreme

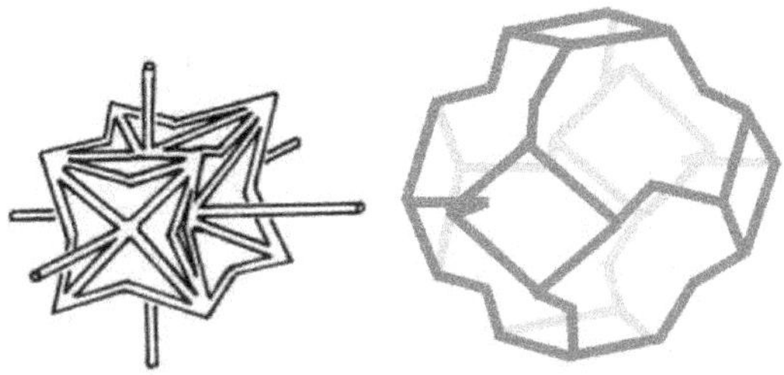

Figure 6.28: Idealized cells for negative Poisson's ratio lattice or foam. Re-entrant cubic cell [60] (left) with permission. Re-entrant tetrakaidecahedron (right).

Most common materials have a Poisson's ratio near 1/3. Also, most foams that are isotropic or nearly so have a Poisson's ratio near 1/3. The allowable range of Poisson's ratio for 3D isotropic materials is from −1 to 0.5. Control of the Poisson's ratio in foams and composites is illustrated by envisaging negative Poisson's ratio foams or lattices which have concave or 're-entrant' cells [60] (Figures 6.29, 6.28) with a negative Poisson's ratio [61]. If one cell has ribs bulging inward, adjacent cells will have ribs bulging outward.

The foams can be isotropic and have a Poisson's ratio as small as −0.7 [60] to −0.8 [62]. These foams are the first 3D isotropic negative Poisson's ratio materials. They are possibly the first elastic metamaterials, though the word had not been used then. Foams can be polymeric (Figure 6.29), metal [62] or ceramic. The Poisson's ratio is governed by the *shape* of the cells, not the rib composition. Negative Poisson's ratio materials have been called "anti-rubber" in a New York Times [63] report on the 1987 Science article. The term auxetic was introduced in a 1991 article [64] that proposed a molecular structure based on a two-dimensional rib structure [13] [65]. Negative Poisson's ratio materials have been called "dilational" [66] in the context of hierarchical laminate structures which easily undergo volume changes but are stiff with respect to shear.

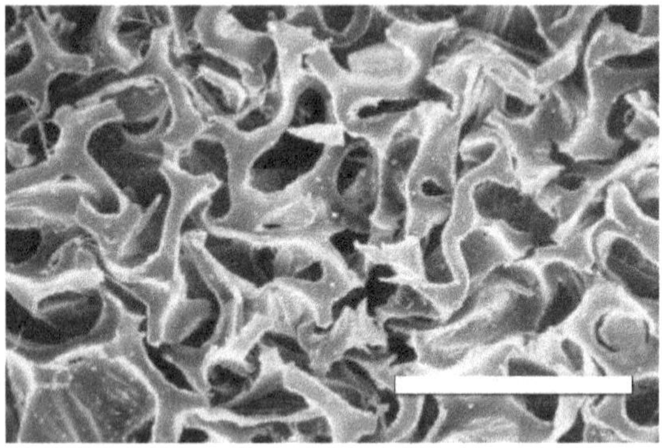

Figure 6.29: Negative Poisson's ratio polymer foam similar to that in [60]. Scale bar: 1 mm.

Poisson's ratio, as with other properties, can be linear or nonlinear. When structural elements such as ribs change angle with deformation, behavior is nonlinear. Honeycombs, when deformed in plane, are nonlinear for that reason [1]. Linear behavior was achieved in a chiral 2D lattice with Pois-

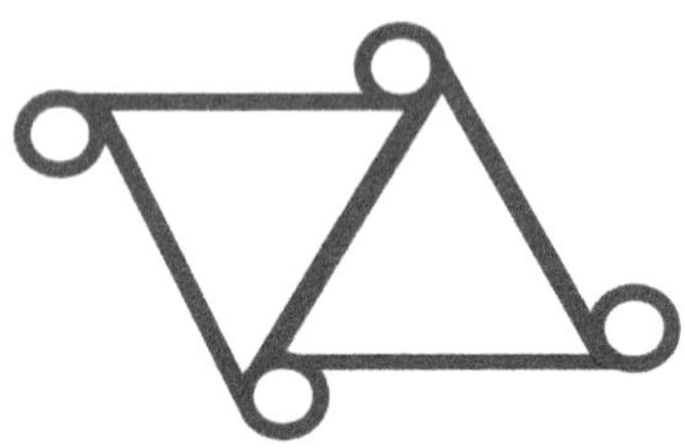

Figure 6.30: Chiral 2D lattice with Poisson's ratio −1, adapted from [67].

son's ratio −1 [67] independent of strain for slender ribs, via analysis and experiment. A structure with rigid rotating squares [68] exhibits a Poisson's ratio of −1 independent of strain, provided the hinges (§6.6.3) at the corners are ideal, frictionless, and stiff with respect to tension.

Control of the Poisson's ratio in cellular solids is done by controlling the cell shape. Polymer foams [60] with a softening point are converted to negative Poisson's ratio foam as follows. The foam is compressed an equal amount in three orthogonal directions by packing it in a mold. The foam is heated to a temperature above the softening point, held at temperature for a period of time that is determined empirically; it is cooled to ambient temperature. Ductile metal foams such as copper foam [69] are converted to negative Poisson's ratio foams as follows. The material is compressed a small amount to permanently deform it in one direction and is then compressed in each orthogonal direction. These permanent compressions are repeated until the change in volume is a factor of three to five. If the foam as received is not perfectly isotropic, the negative Poisson's ratio foam can be made isotropic by modifying the permanent compression procedure to compress more in one direction[70].

It is easier to achieve a negative Poisson's ratio in some directions in anisotropic solids than in all directions in isotropic solids because anisotropy provides more degrees of freedom. A microcellular polymer, polytetrafluoroethylene, if it is expanded in one direction, exhibits a negative Poisson's ratio [71] [72] [73] [74]. These cellular solids are anisotropic because they contain highly oriented fibrils. Poisson's ratio in anisotropic materials can be larger than 1/2 or smaller than −1.

6.6.3 Poisson's ratio tuning in hinged structures: negative or extreme

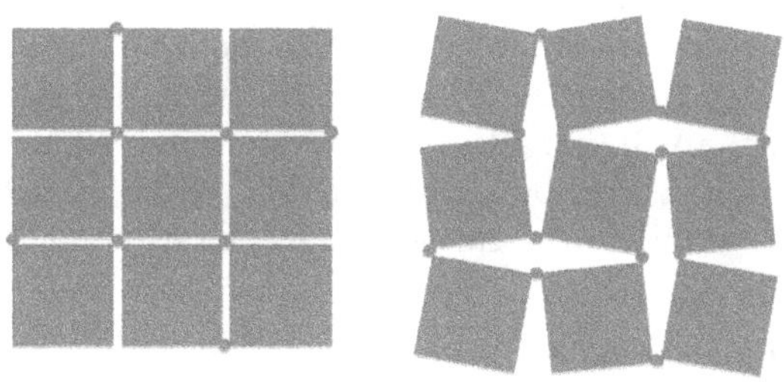

Figure 6.31: Hinged structure with rotating squares with Poisson's ratio −1, adapted from [68]. Hinges are shown as small circles.

Structures containing rigid polygons or polyhedra connected by ideal frictionless hinges have been studied from a mathematical perspective [75] [76] [77] rather than the context of physical properties. They are also called hinged tessellations. These are the earliest known hinged polygon constructions. Such structures have also been studied in the context of physical properties. 2D systems with rotating hexamers [78] [79] give rise to a negative Poisson's ratio. A Poisson's ratio of −1 can occur in 2D structures with rotating rigid squares connected at the corners by ideal hinges [80]; these were obtained via inverse homogenization analyses. Rotating squares were considered [81] as a model for the negative Poisson's ratio in some crystals. A two-dimensional model (Figure 6.31) consisting of rigid rotating squares [68] [82] connected by ideal hinges gave rise to a Poisson's ratio of −1. A hinged array of squares and octagons (Figure 6.32) was

Figure 6.32: Hinged structure with rotating squares and octagons, adapted from [77].

considered as a hinged version [77] of the tessellation shown in Figure 6.2. Rotating hinged triangles can give a similar effect [83]. Related 2D systems with rotating rhombi [84], and prisms [85], also give rise to negative Poisson's ratio. Related structures with hinged components can exhibit negative Poisson's ratio and zero bulk modulus, with arbitrarily large volumetric strain [86]. A 3D structure with sliders as well as hinges [87] was predicted to exhibit a Poisson's ratio of -1. A 3D structure containing hinged cubes [88] exhibits a negative Poisson's ratio and also nonclassical elastic behavior §8.4.5.

Extremal 3D materials [89] have been classified according to the eigenvalues of the elasticity matrix [90], Chapter 30. The number of very small eigenvalues determines the category: unimode if there is one small eigenvalue, bimode if there are two, and so on. A solid with a Poisson's ratio approaching the isotropic lower limit -1 is described as unimode according to this system; there is only one easy mode of deformation, a volume change. A gel or a soft rubber is considered to be pentamode because it can easily deform in five orientations of shear deformation, but it resists hydrostatic deformation. The Poisson's ratio approaches the isotropic upper limit 0.5. For example, a soft rubber may have a shear modulus G of 0.2 MPa and a bulk modulus B of 2 GPa. The Poisson's ratio ν is given for isotropic elastic solids by $\nu = \frac{3B-2G}{6B+2G}$ so for this rubber, $\nu = 0.49995$. A lattice of perfectly hinged linkages in which four ribs meet at a point in a diamond structure is an example of a pentamode material [89] based on the designed structure as shown in Figure 6.33. If the hinges are perfect, $G = 0$ and the Poisson's ratio is exactly 0.5. A diamond crystal, by contrast, has a much lower Poisson's ratio because the interatomic bonds, unlike a pivot, resist rotation.

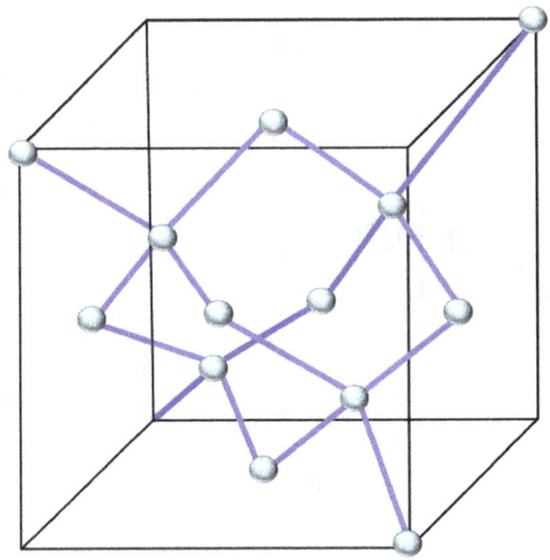

Figure 6.33: Hinged structure with Poisson's ratio approaching 0.5, adapted from [89]. Pivots are shown as spheres.

6.6.4 Poisson's ratio tuning in lattices: negative or extreme

The conceptual visualization of the causal aspect of a negative Poisson's ratio was a lattice cell (Figure 6.28). As with the foams, the Poisson's ratio of such a lattice will depend on strain for sufficiently large strain, giving rise to nonlinearity. As noted above, linear behavior independent of strain for slender ribs was achieved analytically and experimentally in a chiral 2D lattice with Poisson's ratio -1 [67]. Lattices may also be envisaged as hierarchical laminates [66]. The laminates can be elastically isotropic if the lamina thickness and angles are chosen appropriately. In the laminates Poisson's ratio can approach the lower limit -1 if the material represented by the shaded regions in Figure 6.34 is made very soft. The ratio of constituent moduli must be at least a factor 20 to achieve a negative Poisson's ratio. In the limit as one phase becomes void space, then the structure becomes a lattice. The figure shows a two-dimensional laminate; three-dimensional laminates have also been presented.

Compliant 2D micro-mechanisms with a negative Poisson's ratio were designed with the aid of topology optimization and were fabricated using an additive manufacturing (rapid prototyping) method [91]; the cell size

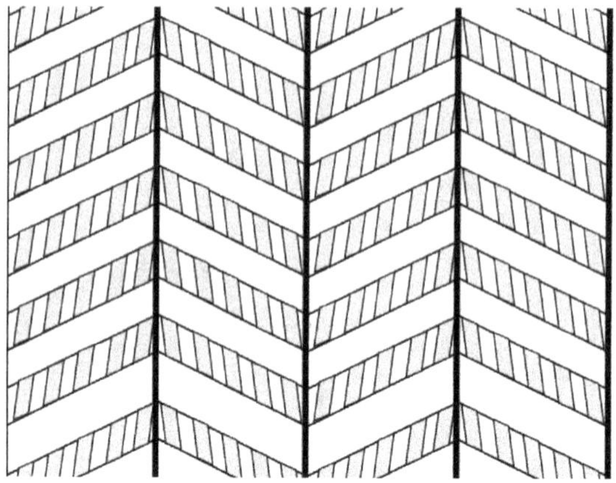

Figure 6.34: Hierarchical laminate with negative Poisson's ratio, adapted from [66]. The structure becomes a lattice if the modulus one phase is zero, corresponding to empty space.

was less than 50μm. Negative Poisson's ratio lattices have been designed [92] based on a rotating square paradigm (Figure 6.31) and made via 3D printing, Figure 6.35. The structure was first designed and represented in digital form. The macroscopic samples in [92] were fabricated using a Objet30 3D printer (Objet (now Stratasys), USA). The basic polymer ink 'FullCure850 VeroGray' was used. A support material was also needed; this was later etched out using NaOH solution. The conceptual basis is an array of tilting squares attached by thin ligaments. The structure was inspired by a two-dimensional model with hinged rigid rotating squares [68] connected by ideal hinges that gave rise to a Poisson's ratio of −1. The symmetry of the 3D lattice is cubic. A Poisson's ratio down to −0.8 was attained, equal to that originally obtained [62] in re-entrant copper foam.

As for zero Poisson's ratio, a Poisson's ratio near zero (0.05) is found in beryllium [93]; the Poisson's ratio can be tuned to zero by control of compression ratio in re-entrant foams [60] [62]. A 2D lattice was designed to have a constant modulus and a Poisson's ratio of zero over a range of strain [94]. The design makes use of curved leaf springs in a planar lattice. Properties are isotropic and linear up to compressive and tensile strains of 0.12. A prototype was made.

Lattices with Poisson's ratios approaching +0.5 based on a hinged pentamode structure [89] [90] were made. As with other hinged structures, it is possible to approximate hinges as slender flexible links or point-like contacts via 3D printing [95]. There are limits to such an approach because the structure becomes fragile if the links are too small.

Chiral 3D lattices [96] have been designed and made via 3D printing (Figure 6.36). The symmetry is cubic, but they can be made elastically isotropic by control of the geometry [97]. There is a size effect in the Poisson's ratio as anticipated in view of the nonzero characteristic length scale in the material as a Cosserat solid (§8.2).

Figure 6.35: Cubic lattice [92] with negative Poisson's ratio based on rotating square concept [68], Figure 6.31, with permission. Scale bar, 20 mm.

Figure 6.36: Cubic chiral lattice with negative Poisson's ratio, designed in [97]. Scale bar: 1 cm.

One may also control the time dependence of Poisson's ratio. This is a form of viscoelastic response §9.3.11. By choosing the time dependence of constituents, one can tune the Poisson's ratio in the time or frequency domain.

Control of mechanical response via design of lattice materials has in recent years become a popular topic as has been reviewed [98].

Applications, Poisson's ratio

Buffers and pads as layers are used to minimize impact force by their compliance. The Poisson's ratio affects the performance of such layers. If the layer is constrained between two stiffer materials, its stiffness is closer to the C modulus than to the Young's modulus E. For a rubbery material with Poisson's ratio approaching 1/2, that is stiff indeed. Therefore wrestling mats are made of foam (with a Poisson's ratio near 1/3), not solid rubber. Buffers that need to be made of solid rubber are textured with grooves to allow for Poisson expansion under compression. Shoe insoles have been used, based on solid grooved rubber or on foam, to reduce impact force in running, with the aim of reducing injury.

6.7 Tuning coupled fields

6.7.1 Tuning thermal expansion: negative or extreme

The thermal expansion of homogeneous materials is a consequence of the slight nonlinearity, also called anharmonicity, of the interatomic force. It had not been considered to be subject to modification. Even in composites, such structures as the Voigt and Reuss structures exhibit thermal expansion that is a weighted average of the expansion values of the constituents. However, as shown below, it is possible to attain extremely high positive or negative expansion, or zero expansion, in heterogeneous materials.

In the derivation of the bounds on thermal expansion (§4.3.4) it is tacitly assumed that the two phases are perfectly bonded, there is no slip, that there is no porosity, and that each phase has a positive definite strain energy. Relaxation of any of these assumptions permits one to achieve arbitrarily large or small values of expansion. This is most easily done via lattices (Figure 6.37) with bi-material ribs [99]. If the spherical nodes in the 3D lattice are considered as pivots, analysis is simplified. Each rib consists of two dissimilar materials bonded together. For such lattice materials one can tune the expansion and the elastic modulus via altering the curvature of the ribs, the aspect ratio or by nesting several lattices. One can achieve thermal expansion that is large positive, zero, or large negative by lattices with curved bi-material ribs [99] as shown in Figure 6.38. The curves shown in the graph for honeycomb and foam are for a steel-Invar based rib; the angle subtended by the curved rib is one radian. Zero expansion lattices based on this concept have been designed [100] with

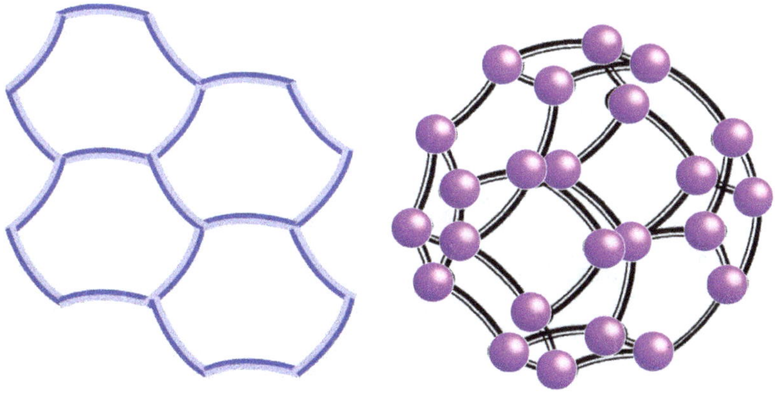

Figure 6.37: Materials with bi-material ribs, adapted from [99]: left, 2D honeycomb; right, 3D lattice cell.

improved ratios of stiffness to density [101] [27]. Given these concepts, one may develop three-phase bounds which allow void space and accommodate the behavior of designed lattices and foams, and which also assume positive definite strain energy.

Zero expansion lattices have been presented [102] based on cells with two kinds of ribs each of which is made with a different material. Zero expansion lattices were designed [103] based on curved, hexagonal honeycomb cells containing triangular inserts made of a second dissimilar material. The triangular element may be comprised of ribs (black) as shown in Figure 6.39. The connecting ribs (blue) are made of a different material. Both materials can have a positive expansion. This type of lattice, because it is stretch dominated, is stiffer than a zero expansion lattice made of curved bend dominated bi-material ribs. In experiments, aluminum and titanium alloys were used; the triangular portion was solid [102]. As with bi-material lattices, one can envisage three-dimensional lattices with ribs made of two different materials. Temperature changes cause rotation at the joints. If the joints are ideal hinges, expansion occurs without internal stress but if the joints are bonded, there is a stress concentration at the joints. The tuning of thermal expansion was done in lattices [104] with star-shaped cells resembling conceptual models used for negative Poisson's ratio [60] in Figure 6.28; the objective was to optimize the behavior.

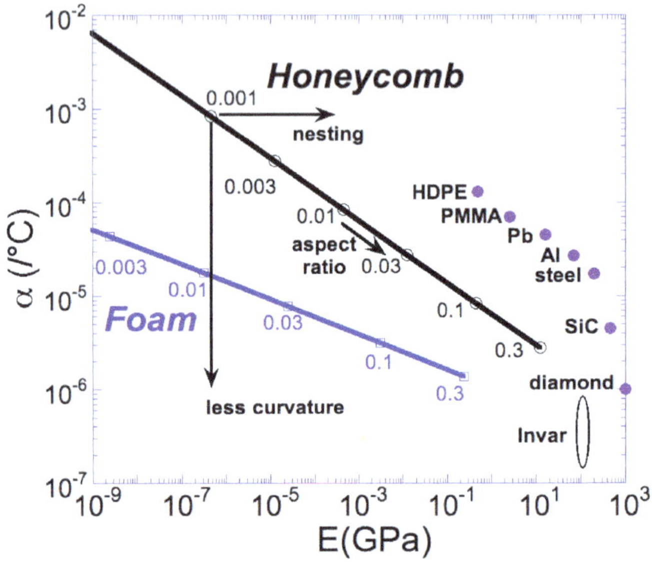

Figure 6.38: Thermal expansion-modulus map of various materials and of designed bi-material honeycomb and foam type lattice materials. Numbers for honeycomb and foam refer to aspect ratio t/L of ribs or cell walls. Adapted from [99].

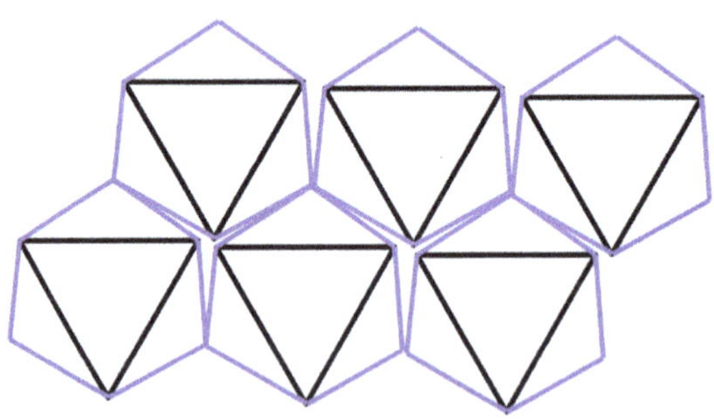

Figure 6.39: Zero expansion lattice, adapted from [102].

6.7.2 Piezoelectric lattices

As with lattices that exhibit controlled thermal expansion, piezoelectric lattices [105] [105] can be made with bi-material ribs. These lattices can exhibit a piezoelectric sensitivity d much larger in magnitude than that of the material comprising the lattice ribs. The ribs bend in response to electric input, giving rise to a much larger displacement than for axial elements. The bending of the ribs, combined with the low overall density, reduces the lattice stiffness considerably in comparison with the solid material comprising the ribs.

6.7.3 Tuning the Hall effect

The Hall effect [107] refers to the action of a magnetic field upon a material carrying an electric current (§4.5). Charge carriers are deflected transversely by the magnetic field to produce a transverse voltage. The sign of this voltage determines the sign of the charge carrier of charge and its magnitude determines the number n of charge carriers of magnitude q per volume [108] [109] provided the material is homogeneous. The Hall coefficient R_H in terms of current density j, applied magnetic field B, and resulting electric field $\mathcal{E}_y$ is $R_H = \frac{\mathcal{E}_y}{j_x B_z}$. This may be expressed $R_H = \frac{1}{nq}$. The Hall effect is used in the design of practical magnetic sensors. It is, however, possible to envisage heterogeneous structures in which the sign of the Hall coefficient is reversed in 2D [110] and in 3D [111]. The structure, consisting of rings, resembles chain mail used in ancient armor. The

Figure 6.40: Cubic chain mail allowing tunable Hall effect [111]. Structure from [112], with permission.

concept was demonstrated experimentally [113] in a structure made via additive manufacturing.

6.8 Control of waves

6.8.1 Role of resonance

The role of resonance in achieving negative or extreme properties at sufficiently high frequency can be traced to the elementary differential equation for a dynamical system with one degree of freedom:

$$m\frac{d^2x}{dt^2} + b\frac{dx}{dt} + kx = F(t).$$

Here m represents an inertia, b is an energy dissipation parameter and k governs a restoring force. In the context of the familiar mass–spring system, m is mass and k is the spring constant. In the context of a resonant electric circuit, m represents an inductance and $1/k$ represents a capacitance.

Suppose the forcing function is $F(t) = Asin2\pi ft$ with f as frequency and A as amplitude. If the damping is small, $b \to 0$, then the compliance x/A becomes large. For frequency f greater than the natural frequency, the phase angle between the cause F and the response x tends to 180°. This can be expressed as a negative compliance x/A. So an array of such resonators can give rise to an effective continuum with properties that take on negative or extreme values over part of the frequency range. The resonators may be mechanical or electromagnetic.

6.8.2 Tuning refraction of waves. Negative index.

The original prospect of negative refractive index was presented by V. G. Veselago [114] in 1968; consequences of negative electric and magnetic properties in refraction of lenses were explored; unusual lens behavior was found. Such materials have negative dielectric permittivity and and magnetic permeability. Veselago recognized at the time that such behavior required dispersion (frequency dependence) of waves and that the effects could be realized in plasmas and in some magnetic materials.

Electromagnetic lattices developed in 1954 and 1962 consisted of a three-dimensional periodic array of intersecting straight wires [115] [116]. For waves with wavelength sufficiently long compared with the lattice spacing, the array behaves as an effective medium with properties similar to those of neutral plasma. Such lattices were used to interpret the propagation of microwaves in the ionosphere. Plasma exhibits dispersion of waves and a characteristic resonant frequency called the plasma frequency.

The resonance of microstructure in heterogeneous materials has been shown to result in extreme dielectric properties, even if the inclusions have a small concentration [117]. It was predicted that negative refractive index can be used to make improved lenses [118] with resolutions substantially superior to the usual diffraction limit that constrains the sharpness of images in the optical microscope. It has been found that heterogeneous periodic structures can exhibit negative refraction [119]. It was found that a periodic array of electrically conductive split ring resonators [120] can exhibit simultaneously negative electric permittivity and magnetic permeability over a range of microwave frequencies near 5 GHz.

A negative refractive index enables waves to be focused with a flat lens. Experiments [121] were done with a periodic array of metal rings and wires based on [119]. The array consisted of 3 by 20 by 20 unit cells as shown in Figure 6.41; the dimensions of the array were 10 by 100 by 100 mm. The flat lens focused diverging microwaves to a point, but high attenuation was observed. In contrast to a curved lens of positive index, the focal length of the negative index lens could be changed by varying the distance between lens and source.

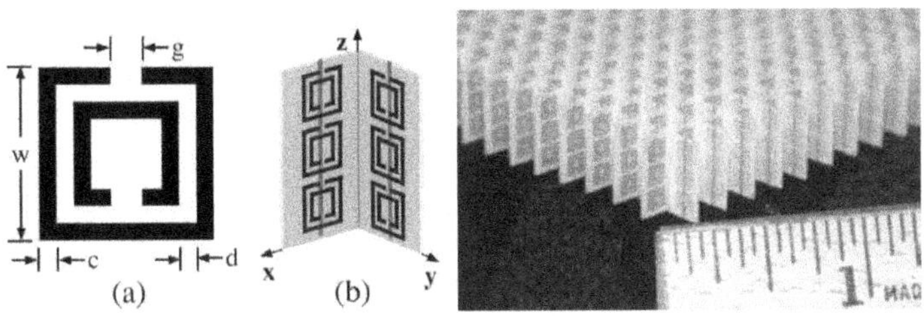

Figure 6.41: Left: Design of ring resonators [120], with permission. Right: Array of ring resonators with negative refractive index acts as a lens for microwaves [121], with permission. Scale ruler shown is in inches.

Negative refraction was observed experimentally in a lattice [122]. This lattice contained square copper split ring resonators and copper wire strips on a substrate of fiber glass: a circuit board. The circuit board segments were cut and assembled into an interlocking lattice.

Chiral structures have been found to be useful in achieving negative refraction [123]. Localized resonance can lead to cloaking: invisibility of objects within wave fields of a particular frequency [124]. Heterogeneous solids which exhibit localized resonance can exhibit negative dielectric or magnetic properties over a narrow band of frequency [125]. Super lenses have been made to focus visible light to a resolution of 70 nm [126], about 1/7 of the wavelength of the light used. This resolution is superior to that of a standard optical microscope. The material used consisted of alternating layers of positive and negative refractive index. The negative index layers were made by depositing structures with fine scale concentric rings of polymer on a gold thin film. Negative refraction of visible light has also been achieved [127] via thin layer waveguides that constrain the light waves.

Quasicrystal lattices

Rib lattices with quasicrystal structure (§3.3.4) were prepared with polymer ribs 1 cm long [128]. The solid volume fraction was about 0.17. Wave transmission in the microwave spectrum (8 GHz to 42 GHz) was measured vs. angle. Wave transmission exhibited a band structure in which regions of strong and weak transmission formed bands. These lattices exhibited band gap properties that were more symmetric than is possible with fully periodic lattices.

6.8.3 Electromagnetic lattices; cloaking

The concept of cloaking objects so that they are invisible to electromagnetic or acoustic waves is a topic of recent interest. Many examples involve coating the object with a material with unusual properties over a range of frequency. The original concept dealt with objects rendered invisible to electromagnetic waves using coated ellipsoids of dielectric material [129]. Local resonances within the material can enable cloaking behavior.

A cloak to hide a copper cylinder was constructed with an array of split-ring resonators, designed for operation over a band of microwave frequencies [130] centered at 8.5 GHz. The structural features were a few millimeters in size. The design process makes use of a coordinate transformation that deforms the space around the object to be cloaked. Resonance in the split rings gives rise to the desired effective dielectric and magnetic properties specified by the design.

6.8.4 Acoustic lattices; cloaking

Metal lattices have been made that mimic the acoustic properties of water, and can be modified to display anisotropic elastic properties suitable for cloaking [131]. The design uses a 2D hexagonal unit cell containing thin load bearing struts, with masses at the vertices to tune the density to be equal to that of water [131] [132]. The width of the interconnecting arms is $\ell = 0.5$ mm, the width of the star arms is $\ell' = 1.51$ mm and the length is h =3.02 mm, and the sides of the regular hexagon are $a = 6.445$ mm. The metal is aluminum and the void space is air. The volume fraction of aluminum is about 0.37. The resulting quasi-static properties are density 1000 kg/m^3, bulk modulus 2.25 GPa and shear modulus 0.065 GPa. These properties are similar to those of water (for which the shear modulus is zero). At frequencies well below structural resonance frequencies in the lattice, the lattice will have acoustic properties similar to those of water. The wave velocity, which depends on the bulk modulus and density, will be the same as that of water. The acoustic impedance, which governs the reflection of waves, will also be the same as that of water. Therefore at low frequency the lattice will be invisible to acoustic wave probes. The low shear modulus causes shear vibration modes to occur at relatively low frequencies. Such modes do not affect acoustic wave propagation significantly at higher frequencies. The lattice has a negative index of refraction over a range of frequency as a result of the resonance of the structural elements. Specifically a branch in the band structure with a negative slope is observed in the frequency range of 70 to 80 kHz. This lattice has a resonant mode structure that allows a slab of the lattice to focus sound [132] at a frequency of 71.2 kHz. Resonant frequencies depend on microstructure size so if the structural elements are made smaller, effects that depend on frequency will occur at higher frequency.

Acoustic cloaking is facilitated if the density at a frequency is matrix valued rather than a scalar. Local resonance can cause the effective mass to appear negative at certain frequencies [133] [134].

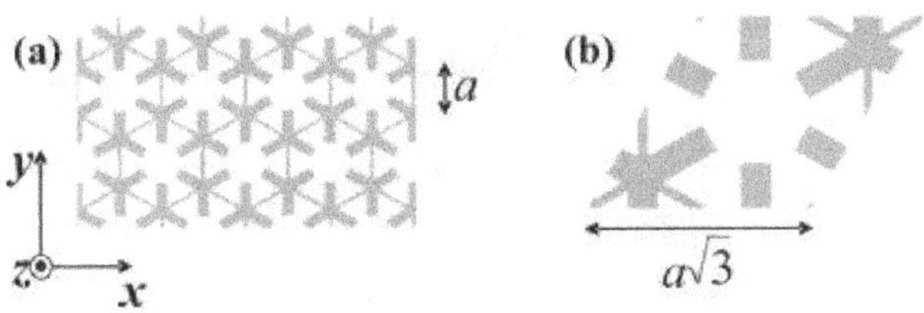

Figure 6.42: Lattice with static and low frequency properties similar to those of water [132], with permission.

6.9 Applications of cellular solids

6.9.1 Applications of foam and honeycomb

Polymer foams are used in cushioning as in seat cushions and mattresses, in packaging to prevent damage to the contents, and as thermal insulators. They are also used in helmets and other protective gear to absorb impact energy by controlled crushing. Polymer foams are used for structural purposes, as cores for sandwich panels in lightweight structures such as those in aircraft. Closed-cell polymer foams are used in marine and other applications in which buoyancy is essential. Syntactic foams provide buoyancy under high pressure because the spherical hollow glass micro-spheres resist collapse; these materials are used in submarine applications.

Metal foams are used as heat exchangers; copper foams can provide more than twice the cooling effectiveness compared with a traditional fin heat exchanger. Such heat exchangers are used in high-performance applications such as medical devices, defense systems, industrial power generation plants, semiconductor manufacturing and in aerospace. Metal foams are also used in air-oil separators and breather plugs in which permeability and resistance to oil are essential. They are also used in high-performance impact energy absorbers in aerospace and military applications under severe environmental conditions. Aluminum is the most commonly used metal in these applications. Ceramic foams based on silicon carbide are used in high temperature heat exchangers.

Honeycombs are used in sandwich structure in aircraft and other applications in which light weight must be combined with stiffness and strength. Cardboard sandwich structure is used in protective packaging.

6.9.2 Applications of lattices

Figure 6.43: Damaged car bumper with simple lattice on a scale of several inches.

As an example of a simple lattice, consider the car bumper shown in Figure 6.43. The lattice is intended to absorb energy during a bœmp. Damage has resulted in the loss of the top cover. In the absence of damage, the lattice structure would not be observed.

Recently, polymer lattices have been made by 3D printing and incorporated in the sole of athletic shoes as shown in Figure 6.44. The lattices are clearly visible, in contrast to lattices within car bumpers. In these examples, lattice ribs at an oblique angle provide compliance as well as a range of deformation that can occur at moderate stress.

Figure 6.44: Lattices in sneakers [135][136], with permission.

6.10 Summary

Cellular solids contain void space and a solid phase. Honeycombs have structural organization in two dimensions; foams have structural organization in three dimensions. If the structure is spatially repetitive, the material is called a lattice. Void space provides design freedom that enables low density materials with high values of ratios of stiffness to weight or strength to weight.

Control of the shape of void space and of the constitution of the solid phase has enabled development of materials with negative, tunable, or extreme properties. These properties include Poisson's ratio, thermal expansion, compressive strength, piezoelectric sensitivity and control of waves. Concepts developed initially in the context of cellular solids have stimulated advances in extreme materials in other contexts.

Bibliography

[1] L. J. Gibson and M. F. Ashby, *Cellular Solids*, Pergamon, Oxford, 1988; 2nd Ed., Cambridge University press, Cambridge, (1997).

[2] B. Grunbaum, G. C. Shephard, Tilings by regular polygons, Math. Mag. 50(5), 227-247 (1977).

[3] K. Critchlow, *Order in Space: A Design Source Book.* New York: Viking Press, (1970).

[4] R. Williams, *The Geometrical Foundation of Natural Structure: A Source Book of Design.* New York: Dover, pp. 35-43, (1979).

[5] H. Steinhaus, *Mathematical Snapshots*, 3rd ed. New York: Dover, pp. 185-190, (1999).

[6] M. Gardner, *The Sixth Book of Mathematical Games from Scientific American.* Chicago, IL: University of Chicago Press, (1984).

[7] `https://en.wikipedia.org/wiki/Cube`

[8] `https://en.wikipedia.org/wiki/Rhombic_dodecahedron`

[9] `https://en.wikipedia.org/wiki/Truncated_octahedron`

[10] Buckminster Fuller, Octahedral building truss, US Patent 3,354,591, (1967).

[11] S. Torquato and Y. Jiao, Dense packings of the Platonic and Archimedean solids, Nature, 460, 876-879 (2009).

[12] S. Torquato and Y. Jiao, Dense packings of polyhedra: Platonic and Archimedean solids, Phys. Rev. E, 80, 041104 (2009).

[13] L. J. Gibson and M. F. Ashby, G. S. Schajer, and C. I. Robertson, The mechanics of two dimensional cellular solids, Proc. Royal Society London, A382: 25-42, (1982).

[14] Specimens provided to the author by Hexcel corporation, Stamford, Connecticut.

[15] H. X. Zhu, J. F. Knott, N. J. Mills, Analysis of the elastic properties of open-cell foams with tetrakaidecahedral cells, J. Mech. Phys. Solids, 45, 319-343 (1997).

[16] Lord Kelvin (Sir W. Thompson), On the division of space with minimal partitional area. Philosophical Magazine, 24(151), 503-514 (1887).

[17] B. Moore, T. Jaglinski, D. S. Stone, and R. S. Lakes, On the bulk modulus of open cell foams, Cellular Polymers, 26, 1-10, March (2007).

[18] General Plastics Company, 4910 Burlington Way, Tacoma, WA 98409 `https://www.generalplastics.com/`

[19] Engineered Syntactic Systems, Attleboro, MA. `www.esyntactic.com`

[20] CRG Industries, syntactic foam, `https://archello.com/product/advantic`

[21] G. Gurtner, M. Durand, Stiffest elastic networks, Proc. R. Soc. A: Math. Phys. Eng. Sci., 470, 20130611 (2014).

[22] T. Tancogne-Dejean, D. Mohr, Elastically-isotropic truss lattice materials of reduced plastic anisotropy, Int. J. Solids Struct., 138, 24-39 (2018).

[23] A. N. Norris, Mechanics of elastic networks, Proc. Royal Soc. London A, 470, 20140522, (1990).

[24] J. L. Grenestedt, Effective elastic behavior of some models for 'perfect' cellular solids, International Journal of Solids and Structures, 36, 1471-1501 (1999).

[25] T. Park, W. Hwang and J. Hu, Plastic continuum models for truss lattice materials with cubic symmetry, Journal of Mechanical Science and Technology 24 (3) 657-669, (2010).

[26] D. Reasa, University of Wisconsin, (2019).

[27] J. Lehman and R. S. Lakes, Stiff, strong, zero thermal expansion lattices via material hierarchy, Composite Structures, 107, 654-663 (2014).

[28] V. Deshpande, N. Fleck and M. F. Ashby, Effective properties of the octet truss lattice material, Journal of the Mechanics and Physics of Solids 49, 1747-1769, (2001).

[29] C. Zener, Contributions to the theory of beta-phase alloys, Phys. Rev. 71, 846-851 (1947).

[30] A. Vigliotti, D. Pasini, Stiffness and strength of tridimensional periodic lattices, Comput. Methods Appl. Mech. Engrg. 229-232 27-43 (2012).

[31] F. Tobias, F. Claudio, K. Muamer, G. Peter and W. Martin, Tailored buckling microlattices as reusable light-weight shock absorbers. Adv. Mater. 28, 5865-5870 (2016).

[32] C. S. Ha, M. E. Plesha, R. S. Lakes, Design, fabrication, and analysis of lattice exhibiting energy absorption via snap-through behavior, Materials and Design, 254, 426-437 (2018).

[33] A. J. Jacobsen, W. Barvosa-Carter, S. Nutt, Compression behavior of micro-scale truss structures formed from self-propagating polymer waveguides, Acta Materialia, 55, 6724-6733, (2007).

[34] D. Reasa and R. S. Lakes, University of Wisconsin (2019).

[35] Jun Li, Nitesh Arora, Gongfeng Li, University of Wisconsin (2018).

[36] X. Zheng, H. Lee, T. Weisgraber, M. Shusteff, J. DeOtte, E. Duoss, J. Kuntz, M. Biener, Q. Ge, J. Jackson, S. Kucheyev, N. Fang, C. M. Spadaccini, Ultralight, ultrastiff mechanical metamaterials, Science, 344 (6190), 1373-1377, (2014).

[37] Z. Hashin, and S. Shtrikman, A variational approach to the theory of the elastic behavior of multiphase materials, J. Mech. Phys. Solids, 11, 127-140, (1963).

[38] T. Tancogne-Dejean, M. Diamantopoulou, M. B. Gorji, C. Bonatti, and D. Mohr, 3D Plate-Lattices: An emerging class of low-density metamaterial exhibiting optimal isotropic stiffness, Adv. Mater., 1803334 (2018).

[39] C. Bonatti and D. Mohr, Mechanical performance of additively-manufactured anisotropic and isotropic smooth shell-lattice materials: Simulations and experiments, Journal of the Mechanics and Physics of Solids, 122, 1-26 (2019).

[40] A. H. Schoen, Infinite periodic minimal surfaces without self-intersections. NASA; Springfield, VA: Federal Scientific and Technical Information. TN D-5541 (1970).

[41] A. H. Schoen, Infinite Regular Warped Polyhedra and Infinite Periodic Minimal Surfaces. Amer. Math. Soc., Abstract 658 30, 23 30 (1968).

[42] A. H. Schoen, A Fifth Intersection-Free Infinite Periodic Minimal Surface of Cubic Symmetry. Amer. Math. Soc., Abstract 664, 664 (1969).

[43] A. H. Schoen, Honeycomb core structures of minimal surface tubule sections, US Patent 3663346, (1972).

[44] A. H. Schoen, Honeycomb panels formed of minimal surface periodic tubule layers, US Patent 3663347, (1972).

[45] W. Fischer and K. Koch, On 3-periodic minimal surfaces, Z. Kristallogr. 179, 31-52, (1987).

[46] M. Wohlgemuth, N. Yufa, J. Hoffman, E. L. Thomas, Triply periodic bicontinuous cubic microdomain morphologies by symmetries, Macromolecules, 34, 6063-6089, (2001).
[47] S. C. Kapfer, S. T. Hyde, K. Mecke, C. H. Arns, G. E. Schröder-Turk, Minimal surface scaffold designs for tissue engineering, Biomaterials, 32, 6875-6882, (2011).
[48] F. P. W. Melchels, K. Bertoldi, R. Gabbrielli, A. H. Velders, J. Feijen, D. W. Grijpma, Mathematically defined tissue engineering scaffold architectures prepared by stereolithography. Biomaterials 31(27), 6909-16, (2010).
[49] S. Torquato, S. Hyun, A. Donev, Multifunctional composites: optimizing microstructures for simultaneous transport of heat and electricity. Phys Rev Lett. 89(26), 266601 (2002).
[50] S. N. Khaderi, V. S. Deshpande, N. A. Fleck, The stiffness and strength of the gyroid lattice, International Journal of Solids and Structures, 51, 3866-3877, (2014).
[51] G. W. Milton, Modeling the properties of composites by laminates. in Erickson, J. L., Kinderlehrer, D., Kohn, R., Lions, J. L. (Eds.), *Homogenization and Effective Moduli of Materials and Media.* Springer Verlag, Berlin, 150-175, (1986).
[52] R. S. Lakes, Materials with structural hierarchy, Nature, 361, 511-515, (1993).
[53] L. R. Meza, A. J. Zelhofer, N. Clarke, A. J. Mateos, D. M. Kochmann, and J. R. Greer, Resilient 3D hierarchical architected metamaterials, PNAS, 112(37) 11502-11507, (2015).
[54] D. Rayneau-Kirkhope, Y. Mao, and R. Farr, Ultralight fractal structures from hollow tubes, Phys. Rev. Lett., 109, 204301, (2012).
[55] T. A. Schaedler, A. J. Jacobsen, A. Torrents, A. E. Sorensen, J. Lian, J. R. Greer, L. Valdevit, W. B. Carter, Ultralight Metallic Microlattices, Science, 334 962-965 (2011).
[56] T. A. Schaedler, private communication (2019).
[57] A. E. H. Love, *A Treatise on the Mathematical Theory of Elasticity*, 4th ed, Dover, New York (1944).
[58] N. Bettenbouche, G. A. Saunders, Lambson, E. F. and Hönle, W. The dependence of the elastic stiffness moduli and the Poisson ratio of natural iron pyrites FeS2 upon pressure and temperature, J. Phys. D., Appl. Phys. 22, 670-675 (1989).
[59] D. J. Gunton and G. A. Saunders, The Young's modulus and Poisson's ratio of arsenic, antimony, and bismuth, J. Materials Science, 7, 1061-1068 (1972).
[60] R. S. Lakes, Foam structures with a negative Poisson's ratio, Science, 235, 1038-1040, (1987).

[61] J. B. Choi and R. S. Lakes, Nonlinear analysis of the Poisson's ratio of negative Poisson's ratio foams, J. Composite Materials, 29, (1),113-128, (1995).
[62] J. B. Choi and R. S. Lakes, Nonlinear properties of metallic cellular materials with a negative Poisson's ratio, J. Materials Science, 27, 5373-5381 (1992).
[63] J. Glieck, Anti-rubber, The New York Times, 14 April (1987).
[64] K. E., Evans, M. A. Nkansah, I. J. Hutchinson, and S. C. Rogers, Molecular network design, Nature, 353, 12, (1991).
[65] A. G. Kolpakov, On the determination of the averaged moduli of elastic gridworks, Prikl. Mat. Mekh, 59, 969-977 (1985)
[66] G. W. Milton, Composite materials with Poisson's ratios close to −1, J. Mech. Phys. Solids, 40, 1105-1137, (1992).
[67] D. Prall and R. S. Lakes, Properties of a chiral honeycomb with a Poisson's ratio −1, Int. J. of Mechanical Sciences, 39, 305-314 (1996).
[68] J. N. Grima and K. E, Evans, Auxetic behavior from rotating squares. J. Mater. Sci. Lett. 19 1563-1565 (2000).
[69] E. A. Friis, R. S. Lakes, and J. B. Park, Negative Poisson's ratio polymeric and metallic foams, Journal of Materials Science, 23, 4406-4414 (1988).
[70] D. Li, Dong, L, and R. S. Lakes, The properties of copper foams with negative Poisson's ratio via resonant ultrasound spectroscopy, Physica Status Solidi (b) 250(10), 1983-1987 (2013).
[71] B. D. Caddock. and K. E. Evans, Microporous materials with negative Poisson's ratio: I. Microstructure and mechanical properties, J. Phys. D., Appl. Phys. 22, 1877-1882 (1989).
[72] K. E. Evans and B. D. Caddock, Microporous materials with negative Poisson's ratio: II. Mechanisms and interpretation, J. Phys. D., Appl. Phys. 22, 1883-1887 (1989).
[73] K. L. Alderson and K. E. Evans, The fabrication of microporous polyethylene having negative Poisson's ratio. Polymer, 33, 4435-4438 (1992).
[74] K. L. Alderson, A. P. Kettle, K. E. Evans, Novel fabrication route for auxetic polyethylene. Part 1. Processing and microstructure, Polymer Engineering and Science, 45, 568-578 (2005).
[75] D. Stuart, Polyhedral and mosaic transformations, Student Publications of the School of Design, North Carolina State University, 12 (1) 2-28 (1963).
[76] D. Wells, *Hidden Connections, Double Meanings*, Cambridge University Press, Cambridge (1988).
[77] G. N. Frederickson, *Hinged Dissections*, Cambridge University Press, Cambridge (2002).

[78] K. W. Wojciechowski, Constant thermodynamic tension Monte Carlo studies of elastic properties of a two-dimensional systems of hard cyclic hexamers, Molecular Physics 61, 1247-125 (1987).
[79] K. W. Wojciechowski, Two-dimensional isotropic system with a negative Poisson ratio, Physics Letters A, 137, 60-64, (1989).
[80] O. Sigmund, Materials with prescribed constitutive parameters: an inverse homogenization approach, Int. J. Solids Struct. 31(17) 2313 - 2329 (1994).
[81] Y. Ishibashi, M. Iwata, A microscopic model of a negative Poisson's ratio in some crystals, Journal of the Physical Society of Japan, 69 2702-2703 (2000).
[82] J. N. Grima, A. Alderson, K. E. Evans, Auxetic behaviour from rotating rigid units. Physica Status Solidi B, 242, 561-75 (2005).
[83] J. N. Grima, and K. E. Evans, Auxetic behavior from rotating triangles, Journal of Materials Science 41, 3193-3196 (2006)
[84] D. Attard and J. N. Grima, Auxetic behaviour from rotating rhombi, Phys. Status Solidi (b) 245, 2395- 2404 (2008)
[85] D. Attard and J. N. Grima, A three-dimensional rotating rigid units network exhibiting negative Poisson's ratios, Phys. Status Solidi B 249, (7), 1330-1338 (2012)
[86] G. W. Milton, Complete characterization of the macroscopic deformations of periodic unimode metamaterials of rigid bars and pivots, J. Mechanics and Physics of Solids, 61, 1543-1560 (2013).
[87] R. F. Almgren, An isotropic three dimensional structure with Poisson's ratio = -1, J. Elasticity, 15, 427-430, (1985).
[88] C. Andrade, Ha, C. S., R. S. Lakes, Extreme Cosserat elastic cube structure with large magnitude of negative Poisson's ratio, Journal of mechanics of materials and structures (JoMMS), 13(1), 93-101 (2018).
[89] G. W. Milton, A.V. Cherkaev, Which elasticity tensors are realizable? ASME J. Eng. Mater. Technol. 117, 483-493 (1995).
[90] G. W. Milton, *The Theory of Composites*, Cambridge University Press, Cambridge (2002).
[91] U. D. Larsen, O. Sigmund, and S. Bouwstra, Design and fabrication of compliant micromechanisms and structures with negative Poisson's ratio, IEEE Journal of Microelectromechanical Systems, 6, 99-106, (1997).
[92] T. Buckmann, R Schittny, M. Thiel, M Kadic, G. W. Milton, and M Wegener, On three-dimensional dilational elastic metamaterials, New Journal of Physics 16 033032 (2014).
[93] A. Migliori, H. Ledbetter, D. J. Thoma, and T. W. Darling, Beryllium's monocrystal and polycrystal elastic constants, J. Applied Physics 95, 2436-2440 (2004).

[94] A. Delissen, G. Radaelli, L. A. Shaw, J. B. Hopkins and J. L. Herder, Design of an isotropic metamaterial with constant stiffness and zero Poisson's ratio over large deformations, J. Mech. Des 140(11), 111405 (10 pages) (2018).
[95] M. Kadic, T. Buckmann, N. Stenger, M. Thiel, and M. Wegener, On the practicability of pentamode mechanical metamaterials, Appl. Phys. Lett. 100, 191901 (2012).
[96] C. S. Ha, M. E. Plesha, R. S. Lakes, Chiral three dimensional lattices with tunable Poisson's ratio, Smart Materials and Structures, 25, 054005 (6pp) (2016).
[97] C. S. Ha, M. E. Plesha, R. S. Lakes, Chiral three-dimensional isotropic lattices with negative Poisson's ratio, Physica Status Solidi B, 253, (7), 1243-1251 (2016).
[98] A. Zadpoor, Mechanical meta-materials, Materials Horizons, Royal Society of Chemistry, 3, 371-381, (2016).
[99] R. S. Lakes, Solids with tunable positive or negative thermal expansion of unbounded magnitude, Applied Phys. Lett. 90, 221905 (2007).
[100] J. Lehman and R. S. Lakes, Stiff lattices with zero thermal expansion. J. Intell. Mater. Syst. Struct., 23(11): 1263-1268 (2012).
[101] J. Lehman and R. S. Lakes, Stiff, strong zero thermal expansion lattices via the Poisson effect, Journal of Materials Research, 29, 2499-2508, (2013).
[102] C. A. Steeves, S. L. dos Santos e Lucato, M. He, E. Antinucci, J. W. Hutchinson, A. G. Evans, Concepts for structurally robust materials that combine low thermal expansion with high stiffness, J. Mech. Phys. Solids, 55, 1803-1822 (2007).
[103] G. Jefferson, T. A. Parthasarathy, R. J. Kerans, Tailorable thermal expansion hybrid structures. Int. J. Solids Struct, 46, 2372-2387 (2009).
[104] J. B. Hopkins, K. J. Lange and C. M. Spadaccini, Designing microstructural architectures with thermally actuated properties using freedom, actuation, and constraint topologies, J. Mech. Des 135(6), 061004 (2013).
[105] R. S. Lakes, Piezoelectric composite lattices with high sensitivity, Philosophical Magazine Letters 94, (1), 37-44 (2014).
[106] B. Rodriguez, H. Kalathur, and R. S., Lakes, A sensitive piezoelectric composite lattice: experiment, Physica Status Solidi, 251(2) 349-353 (2014).
[107] E. Hall, On a new action of the magnet on electric currents, American Journal of Mathematics, 2 (3): 287-292 (1879).
[108] D. Halliday and R. Resnick, *Physics*, vol. II, J. Wiley, New York, (1962).

[109] D. W. Preston and E. R. Dietz, *The Art of Experimental Physics*, J. Wiley, New York, (1991)
[110] M. Briane, D. Manceau, and G. W. Milton, Homogenization of the two-dimensional Hall effect. J. Math. Ana. App., 339(2), 1468-1484 (2008).
[111] M. Briane and G. W. Milton, Homogenization of the three-dimensional Hall effect and change of sign of the Hall coefficient, Archive for Rational Mechanics and Analysis, Springer Verlag, 193 (3), 715-738 (2009).
[112] Dylon Whyte, `http://artofchainmail.com/patterns/japanese/hitoye_gusari.html`
[113] C. Kern, M. Kadic, M. Wegener, Experimental evidence for sign reversal of the Hall coefficient in three- dimensional metamaterials. Phys. Rev. Lett., 118, 016601 (2017).
[114] V. G. Veselago, The electrodynamics of substances with simultaneously negative values of ϵ and μ, Sov. Phys. Uspekhi 10, 509-514, (1968).
[115] R. N. Bracewell, Wireless Engineer, Iliff and Sons Ltd., London, p. 320. (1954).
[116] W. Rotman, IRE Trans. Antennas Propag. AP10, 82 (1962).
[117] N. A. Nicorovici, R. C. Mcphedran, G. W. Milton, Optical and dielectric properties of partially resonant systems, Phys. Rev. B 49, 8479-8482, (1994).
[118] J. B. Pendry, Negative refraction makes a perfect lens, Phys. Rev. Lett. 85, 3966 (2000).
[119] D. R. Smith, W. J. Padilla, D. C. Vier, S. C. Nemat-Nasser, and S. Schultz, Composite medium with simultaneously negative permeability and permittivity, Phys. Rev. Lett. 84, 4184 (2000). Figure reproduced from this article, with the permission of AIP Publishing.
[120] R. A. Shelby, D. R. Smith, S. C. Nemat-Nasser, and S. Schultz, Microwave transmission through a two-dimensional, isotropic, left-handed metamaterial, Appl. Phys. Lett. 78, 489-491 (2001).
[121] J. D. Wilson, Z. D. Schwartz, and C. T. Chevalier, Multifocal flat lens demonstrated with left-handed metamaterial, NASA report, Research and Technology 2005, NASA/TM-2006-214016, E-15297 p. 21-22 (2006).
[122] R. A. Shelby, D. R. Smith, and S. Schultz, Experimental verification of a negative index of refraction, Science 292, 77 (2001).
[123] J. B. Pendry, A chiral route to negative refraction, Science, 306: 1353-1355, (2004).
[124] N. A. Nicorovici and G. W. Milton, On the cloaking effects associated with anomalous localized resonance, Proceedings of the Royal Society of London, A, 462, 3027 - 3059, (2006).

[125] S. Linden, Enkrich, C., Wegener, M., Zhou, J., Koschny, T., and C. M. Soukoulis, Magnetic response of metamaterials at 100 terahertz, Science, 306, 1351-1353, (2004).
[126] I. S. Smolynaninov, Y. J. Hung, C. C. Davis, Magnifying superlens in the visible frequency range, Science, 315, 1699-1701 (2007).
[127] H. J. Lezec, J. A. Dionne, H. A. Atwater, Negative refraction at visible frequencies, Science, 315, 1699-1701 (2007).
[128] W. Man, M. Megens, J. Steinhardt, P. M. Chaikin, Experimental measurement of the photonic properties of icosahedral quasicrystals, Nature, 436, 993-996 (2005).
[129] M. Kerker, Invisible bodies, J. Opt. Soc. Am. 65 376-379 (1975).
[130] D. Schurig, J. J. Mock, B. J. Justice, S. A. Cummer, J. B. Pendry, A. F. Starr, D. R. Smith, Metamaterial electromagnetic cloak at microwave frequencies, Science, 314 (5801), 977-980, (2006).
[131] A. N. Norris, A. J. Nagy, Metal Water: A metamaterial for acoustic cloaking, Phononics 2011: First International Conference on Phononic Crystals, Metamaterials and Optomechanics, Santa Fe, NM pp. 112-113 (2011).
[132] A. C. Hladky-Hennion, J. O. Vasseur, G. Haw, C. Croenne, L. Haumesser, and A. N. Norris, Negative refraction of acoustic waves using a foam-like metallic structure, Appl. Phys. Lett. 102(14), 144103 (2013). Figure reproduced from this article with the permission of AIP Publishing.
[133] G. W. Milton, New metamaterials with macroscopic behavior outside that of continuum elastodynamics, New J. Phys. 9(10), 359-359 (2007).
[134] G. W. Milton and J. R. Willis, On modifications of Newtons second law and linear continuum elastodynamics, Proc. R. Soc. Lond. A 463, 855-80 (2007).
[135] Thanks to Matthew Flail and Timothy Ganter, footprint-footwear.com, (2019).
[136] Thanks to Aarish Netarwala, Design and Strategy (2019).

Chapter 7

Biological material structural hierarchy

7.1 Introduction

Most materials of biological origin exhibit a rich hierarchical structure. Because it is not easy to fully characterize all the constituents, particularly those near interfaces, it can be a challenge to obtain sufficiently complete theoretical analyses of the structure–property relations. Experimental investigation is therefore crucial.

Properties of biological materials are studied in the context of basic science and in support of clinical diagnosis and surgery in which these materials are pertinent. Biological materials are also of interest to the scientist or engineer who seeks to imitate biological materials (to design bio-mimetic materials) or draw inspiration from the rich structure of biological materials (to design bio-inspired materials). The biological materials considered here are bone and teeth, ligament and tendon, and wood.

7.2 Bone and teeth

Bones and teeth are mineralized biological composites. Compact or cortical bone (Figure 7.1) is dense (compared with soft tissue or spongy bone), fibrous and strong; it forms the mid-shaft of long bones and the outer layer of the skull. Spongy bone, also called cancellous or trabecular bone, is porous and of lower density. It occurs at the ends of long bones and in the core of vertebrae. Teeth contain enamel, the outer portion, and dentin, within the enamel, and pulp at the core of the tooth.

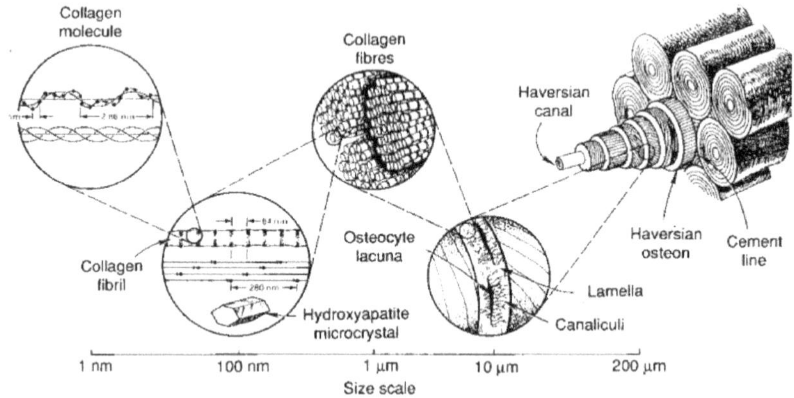

Figure 7.1: Hierarchical structure of human Haversian bone [3], with permission.

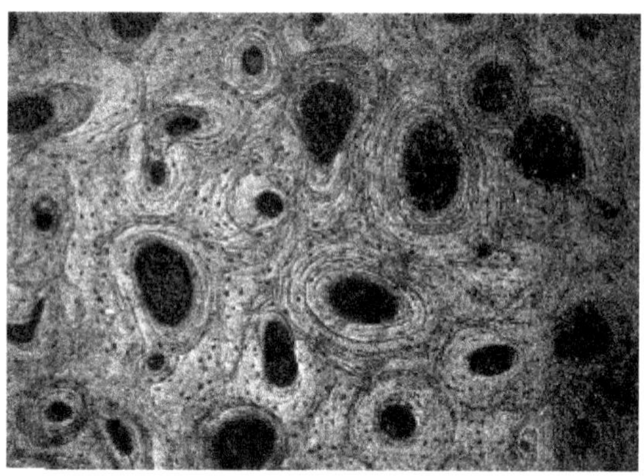

Figure 7.2: Structure of human Haversian bone; cross section of compact bone, [5], with permission. Original magnification, about 67.

7.2.1 Compact bone: stiffness and strength

Human compact bone is a natural composite which exhibits a complex hierarchical structure [1] [2] [3] as shown in Figure 7.1. The large fibers in human compact bone are called osteons; they are about 0.2 mm in diameter and appear as circles or ovals in a cross section micrograph (Figure 7.2). The dark circles and ellipses surrounded by concentric rings in the image are the Haversian canals; the rings are lamellae. Compact bone has a strength to weight ratio greater than that of structural steel; it is also stronger than brick, wood or concrete. Bone as a living material is able to repair

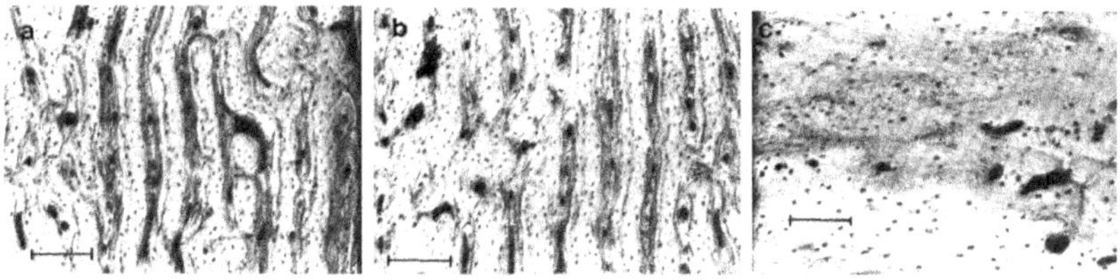

Figure 7.3: Structure of bovine compact plexiform bone [10] with permission; (a) cut perpendicular to longitudinal direction, (b) cut perpendicular to circumferential direction, (c) cut perpendicular to radial direction. Scale mark, 0.2 mm.

microscopic damage; it is therefore not so vulnerable to mechanical fatigue that can cause fracture in engineering materials. Bone is anisotropic and viscoelastic. Bovine compact bone has a different micro-structure, called plexiform; it is laminar rather than fibrous (Figure 7.3); bovine bone is stiffer and stronger than human bone.

Human compact bone is considered to be axisymmetric [5]; bovine compact bone is considered to be orthotropic in symmetry. Elastic modulus E components and strength for bone via quasi-static mechanical tests are compared with tooth dentin and enamel in Table 7.1. Elastic constants of bovine bone were initially evaluated [6] under an assumption of transverse isotropy.

Table 7.1: Properties of compact bone [2] [4] (via static tests) and tooth dentin and enamel: density ρ, Young's modulus E. Strength σ^{ult} is given for tension and compression. L refers to longitudinal, T refers to transverse.

Property	Human Haversian	Cow Haversian	Cow plexiform	Dentin	Enamel
$\rho (kg/m^3)$	1800-2000	2060	-	2.1	2.9
E_L(GPa)	17	22.6	26.5	18	50
E_T(GPa)	11.5	10.2	11.0	-	
G_{LT}(GPa)	3.3	3.6	5.1	-	
$\sigma^{ult}_{L,compr}$(MPa)	193	254	294	138	
$\sigma^{ult}_{T,compr}$(MPa)	133	146	-	-	
$\sigma^{ult}_{L,tens}$(MPa)	148	144	167		
$\sigma^{ult}_{T,tens}$(MPa)	49	46	55		

Elastic modulus C components for bone obtained by ultrasound are given in Table 7.2. The results for dry human bone were obtained at 5 MHz and those for wet bovine bone were obtained at 2.25 MHz. Results [9] for bovine bone were based on tests assuming orthotropic symmetry; the other results [7] [8] assumed transverse isotropy, hence pairs of equal elastic constants. Recall that the C constants differ from the engineering elastic constants. When all the elastic constants are known, it is possible to invert the C matrix to obtain the compliance matrix, hence engineering or technical constants such as Young's modulus E. Because bone is a viscoelastic material, its properties depend on frequency. Moduli at ultrasonic frequencies are higher than moduli at frequencies associated with physical activities of the body or with quasi-static testing.

Table 7.2: Elastic modulus components for bone via ultrasound.

Modulus	Bovine [7]	Bovine [9]	Human [8]
C_{11}(GPa)	19.7	14.1	23.4
C_{22}(GPa)	19.7	18.4	23.4
C_{33}(GPa)	32	25	32.5
C_{44} (GPa)	5.4	7.0	8.71
C_{55} (GPa)	5.4	6.3	8.71
C_{66} (GPa)	3.8	5.28	7.17
C_{12}(GPa)	12.1	6.34	9.06
C_{13}(GPa)	12.6	4.84	9.11
C_{23}(GPa)	12.6	6.94	9.11

Properties of compact bone from different species are given in Figure 7.4. Deer antler has the highest toughness and the lowest stiffness. That is helpful because the animals subject their antlers to repeated impacts. The tympanic bulla (ear bone) of a whale has the highest stiffness and density but the least toughness. The ear bone has an acoustic role in which density and stiffness are beneficial; the ear bone is protected within the skull so it does not undergo impacts. The femur of a cow has intermediate properties.

The mineral phase of bone is crystalline hydroxyapatite ($Ca_{10}(PO_4)_6(OH)_2$) which is virtually elastic; it provides the stiffness of bone [11]. Compact bone contains about 69% mineral by weight. The organic component is mostly the fibrous protein collagen, 22% by weight; there is about 9% water and 1% polysaccharide and related material. On the microstructural level are the osteons [12], which are large ($\sim 200 \mu m$ diameter) hollow fibers composed of concentric lamellae and of pores. The osteons confer toughness upon the bone. Lacunae are ellipsoidal pores with dimensions on the order 10 μm which provide spaces for the osteocytes (bone cells) which maintain the bone and allow it to adapt to changing conditions of stress by mediating growth or resorption of bone in response to stress. Haversian canals contain blood vessels which nourish the tissue, and nerves for sensation. The flow of fluid within the pore space in bone is important in the nutrition of bone cells.

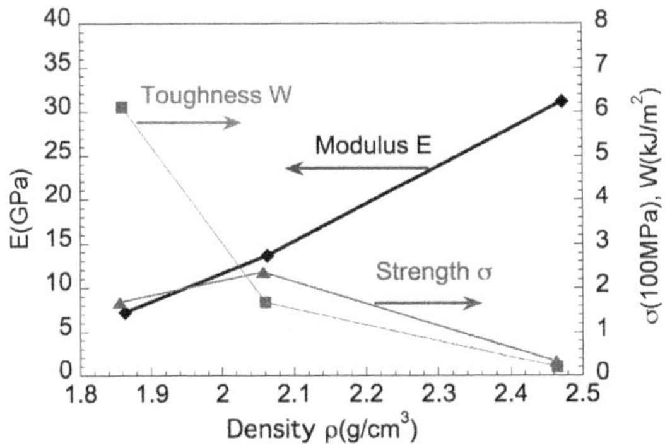

Figure 7.4: Diagram of toughness W, Young's modulus E and strength σ vs. density in g/cc of different kinds of bone: antler, femur, and whale bulla (ear bone), adapted from [2].

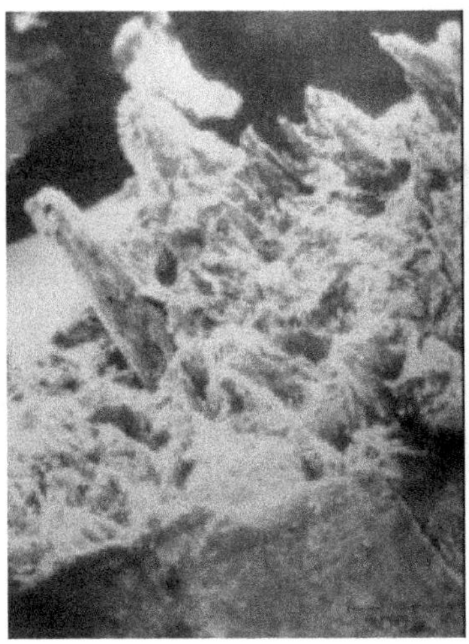

Figure 7.5: Pullout of osteons in a fracture surface of a laboratory bone specimen [14], with permission.

As for toughness, energy is dissipated in the pullout of osteons from the matrix as shown in Figure 7.5, during crack propagation. The lamellae are built of fibers, and the fibers contain fibrils (smaller fibers). At the ultrastructural level (nanoscale) the fibers are a composite of the mineral hydroxyapatite and the protein collagen, which has a triple helix structure. Specific structural features have been associated with properties such as stiffness via the mineral crystallites [1], creep via the cement lines between osteons [13], and toughness via osteon pull-out at the cement lines [14].

As for strain in bones, the maximum strain along the tibia axis of a human volunteer was about 3.5×10^{-4} during walking and 8×10^{-4} during jogging [20]. Strains of similar magnitudes have been observed in animals [21]. Strain in bones of human athletes remains below 2×10^{-3} even during strenuous activity. The largest strain magnitude observed in the normal activity of an animal was 3.2×10^{-3} in the tibia of a galloping horse [22]; in race horses, strain in leg bone may exceed 5×10^{-3}. In comparison, in tension in the longitudinal direction human bone yields at a strain of 6.7×10^{-3} and fractures at a strain of 0.03. The strain levels observed in vivo are significant in view of the fatigue properties of bone. Bone does accumulate micro-cracks associated with fatigue damage. Such cracks are eventually removed during the natural remodeling of bone by the body.

7.2.2 Compact bone: piezoelectricity

Dry bone is piezoelectric (§4.2) in the classic sense, i.e. mechanical stress results in electric polarization, the indirect effect; and an applied electric field causes strain, the converse effect [23]. The magnitude of the piezoelectric sensitivity coefficients of bone depends on frequency, on direction of load, and on relative humidity [24] [25] [26]. Values up to 0.7 pC/N have been observed, to be compared with 0.7 and 2.3 pC/N for different directions in quartz. Bone and tendon are also pyroelectric [27]. The piezoelectric properties of bone are of interest in view of their hypothesized role in stimulating bone remodeling. In wet bone [28] [29] streaming potentials can result in stress generated potentials at relatively low frequencies even in the presence of dielectric relaxation or electrical conductivity.

7.2.3 Compact bone: adaptation

Bone responds to prevailing stress conditions by adding bone or removing bone. The relationship between the mass and form of a bone to the forces applied to it was appreciated by Galileo [30], who is credited with being the first to understand the balance of forces in beam bending and with applying this understanding to the mechanical analysis of bone. Julius Wolff published a seminal 1892 monograph [31] on bone remodeling. The observation that bone is reshaped in response to the forces acting on it is presently referred to as Wolff's law. Immobilization of humans causes loss of bone and excretion of calcium and phosphorus. Long spaceflights under zero gravity also cause loss of bone. Exercise results in addition of bone tissue, and stronger bones. Bone remodeling appears to be governed by a feedback system in which the bone cells sense the state of strain in the bone matrix around them and either add or remove bone as needed to maintain the strain within normal limits. Bone turnover in living organisms also removes tissue that has undergone micro-cracking or other forms of damage. The damaged tissue is replaced with new bone. This repair of damage renders living bone immune to fatigue failure from repeated deformation. An exception occurs if the person increases the level of physical activity more rapidly than the body can adapt; painful cracks or breaks in the bone are called march fractures because they can occur in military recruits. The adaptive response of bone to mechanical stimuli is mediated by living cells. The process or processes by which the cells are able to sense the strain and the important aspects of the strain field are presently unknown. Electrical effects in bone have been considered in a causal role.

Inspired by the hypothetical role of bone electricity in maintenance and growth of bone, external electrical stimulation [32] of bone has been used clinically to potentiate and accelerate healing.

7.2.4 Compact bone: stress concentration

In bone, stress concentrations around holes and cracks are lower than predicted by elasticity theory [33]. The deviation is of larger magnitude for small holes than for larger ones. This is pertinent to the function of bone because micro-cracks occur as a result of fatigue damage, and holes are drilled for screws used in surgical repairs. Stress concentration occurs in the presence of large gradients of strain which cause bending or twisting of the largest structural elements, the osteons. Sensitivity to gradients may be understood in the context of generalized continuum theory as discussed in §8.2. Cavities and channels that occur naturally in bone tend to be oriented to minimize their stress-concentrating effects [34]. Such small cavities also can impede or stop cracks.

7.2.5 Spongy bone

Cancellous or trabecular bone is a highly porous or cellular form of bone. Figure 7.7 shows a photograph of cancellous bone from the leg bone of an animal. In a typical long bone, the cortex or exterior of the shaft (diaphysis) and flared ends (metaphysis) is composed of compact bone while the interior, particularly near the articulating ends, is filled with cancellous bone. Cancellous bone may also be found to fill the interior of short bones and flat bones as well as in the interior of bony tuberosities under muscle attachments. The structure of cancellous bone is that of a latticework of bars and plates. The volume fraction of solid material can be from 5% to 70%; the interstices in living bone are filled with marrow.

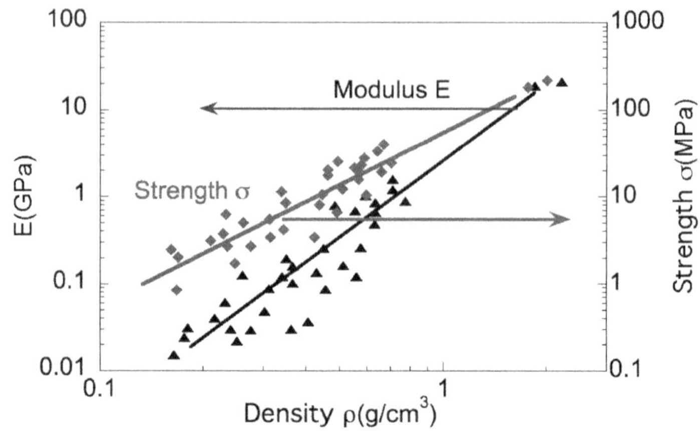

Figure 7.6: Bone properties, Young's modulus E and strength vs. density based on [15] [2] [16].

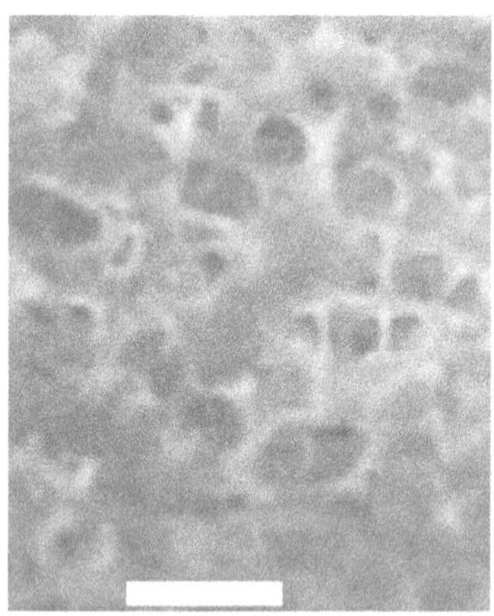

Figure 7.7: Spongy bone; scale bar 2 mm.

The compressive strength σ_{ult} [in MPa] of bone depends very much on the density ρ (in g/cm^3) (Figure 7.6) and also varies with the strain rate $d\epsilon/dt$ as follows [16]. $\sigma_{ult} = 68(d\epsilon/dt)0.06(\rho^2)$. This sort of spongy bone behaves as an open-cell foam. Indeed, the elastic modulus is proportional to the first, second or third power of the density, depending on the detailed structure of the spongy bone. The first power corresponds to the aligned structure of columns as found in vertebrae. The second power corresponds to a typical open structure of rods, and the third power corresponds to a structure of plates, which can occur at higher density.

Density is not, however, the only determinant of the properties of cancellous bone. The microstructure can vary considerably from one part of the body to another. For example, in the vertebrae and in the tibia [17], a highly oriented, columnar architecture is observed. This kind of trabecular bone is highly anisotropic: the Young's modulus in the longitudinal direction can exceed that in the transverse direction by more than a factor of ten [17]. By contrast, in regions such as the proximal part of the bovine humerus, the cancellous bone can be essentially isotropic [18]. This bone is about twice as strong in compression as in tension. In many ways cancellous bone [19] is similar in its behavior to man-made rigid cellular foams. For example, in compression, the stress-strain curves contain a linear elastic region, up to a strain of about 0.05, at which the cell walls bend or compress. A plateau region of almost constant macroscopic stress is associated with elastic buckling, plastic yield, or fracture of the cell walls. The compressive failure of the cancellous bone proceeds at approximately constant stress until the cell walls touch each other; at this point any further compression causes the stress to rise rapidly. By contrast, the fracture of cancellous bone in tension proceeds abruptly and catastrophically [18]. The energy absorption capacity of cancellous bone is consequently much less in tension than it is in compression [19]. This suggests that tensile and avulsion fractures of cancellous bone observed clinically are associated with minimal energy absorption and therefore may be precipitated by relatively minor trauma.

7.2.6 Teeth

Properties of tooth dentin and enamel are compared with those of bone in Table 7.1.

Tooth enamel consists mostly of (about 99%) hydroxyapatite mineral, with a small amount of protein. Dentin is similar to bone in its content of protein and mineral as well as in its mechanical properties, but has a different structure. Dentin has oriented porosity called dentinal tubules.

7.3 Ligaments and tendons

Ligaments are fibrous links that connect bones to each other; they stabilize joints. Tendons connect muscles to bones. Tendons and ligaments contain bundles of oriented collagen fibers. Tendons and ligaments are both anisotropic and viscoelastic (Chapter 9). Tendons (Figure 7.8), similarly to collagenous ligaments, have a hierarchical structure [35]. Fibers are organized in a zig-zag wavy structure at times referred to as crimp (Figures 7.8, 7.9).

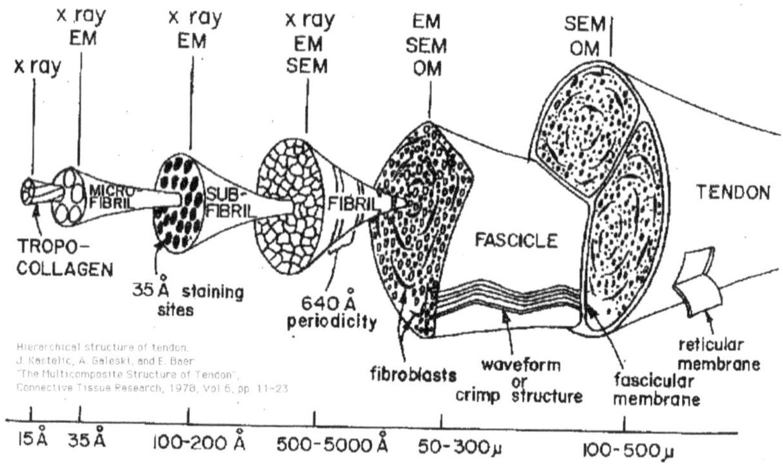

Figure 7.8: Hierarchical structure of tendon [35] with permission.

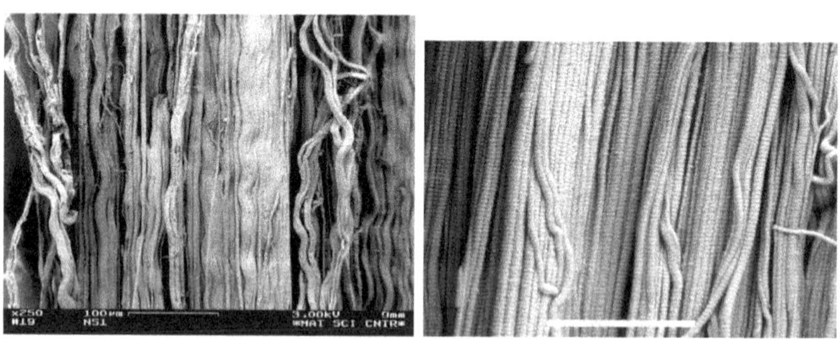

Figure 7.9: Ligament structure, rat [36], with permission. Left, scale bar, 100 μm. Right, higher magnification, scale bar, 2 μm

Tendons and ligaments exhibit a strain-stiffening nonlinearity [37] in which the slope of the stress-strain curve increases with strain; this is observed in animal and in human tendons and ligaments. The strain-stiffening type nonlinearity of ligament allows free motion of a joint in normal activities with minimal effort required to stretch the ligament. If a joint is subject to excess load, the ligaments become stiffer by virtue of their nonlinearity, to provide stability. The nonlinearity is manifest in stress-strain curves for ligaments that are concave up (Figure 7.10). The strain-stiffening nonlinearity is attributed to a straightening of initially wavy collagen fibers in the ligament (Figure 7.9). Ligament strain during normal bodily activity is typically below 4% [38]; from 3.6% for squatting, to 1.7% for bicycling [39]. Structural damage in laboratory specimens begins to be manifest as laxity upon recovery following strain exceeding 5.1% [40]; cellular damage is initially observed following strain below 1% and increases linearly with strain. Sub-failure damage in ligaments results in broken and disordered fibers while other fibers are intact. A ligament sprain in a living person is painful, but some cellular damage may occur before pain is perceived.

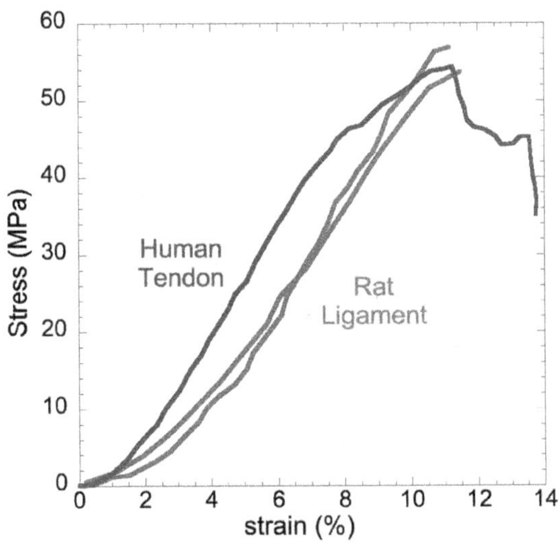

Figure 7.10: Stress vs. strain is nonlinear for left and right rat knee ligaments, adapted from [40] and for human patellar tendons adapted from [42]. Normal activities result in strain up to about 4%.

Tendon properties have been explored in living human subjects by measuring the motion of bony prominences by ultrasound [44] and by measuring the force exerted by a limb. Nonlinear force-deformation characteristics, similar to those for laboratory specimens, are observed; for maximum muscle tension, the stress is 25 MPa, the strain is 0.025 and the effective modulus at that strain is 1.2 GPa. Therefore the tendon usually operates in the toe region (the initial concave-up part of the curve) of its nonlinearity.

7.4 Wood and other plant tissue

7.4.1 Wood structure, stiffness and strength

Wood is well established as a material used in structures. The ratio $\frac{E}{\rho}$ of elastic modulus to density for wood is similar to that of metals such as steel and aluminum. The figure of merit $\frac{E}{\rho^2}$ for beam bending and for plate bending $\frac{E}{\rho^2}$ as discussed in §1.3 favors materials of low density. In that context, as a result of the low density, wood has substantially higher figures of merit for bending than structural metals.

The properties of wood are strongly dependent on density as shown in Figure 7.11. Dry oak wood [45] of 12 percent moisture content has a density of about 0.6 g/cc and a longitudinal Young's modulus of 11.3 GPa. The properties along the grain are stretch-dominated as expected from the grain orientation, while the properties across the grain are bend-dominated. This contrast is associated with the large anisotropy of wood. The strength

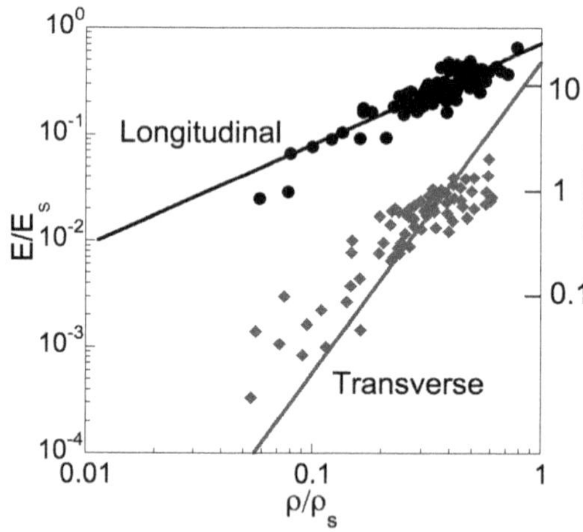

Figure 7.11: Young's modulus E and normalized Young's modulus of wood vs. relative density, adapted from [15]. The solid line for the longitudinal direction is based on $\frac{E}{E_s} = C_L \frac{\rho}{\rho_s}$. The solid line for the transverse directions is based on $\frac{E}{E_s} = C_T [\frac{\rho}{\rho_s}]^3$. E_s is the modulus of the solid constituent.

of wood in the longitudinal direction is proportional to the relative density $\frac{\rho}{\rho_s}$. A dense wood of density 1 g/cm^3 will have a longitudinal compressive strength of about 100 MPa. The strength of wood in a transverse direction is proportional to the square of the relative density: $[\frac{\rho}{\rho_s}]^2$. At all density values, wood strength is considerably less transversely than longitudinally.

Plant tissues such as wood [47] are cellular solids with complex hierarchical structures. For simplicity, wood construction products are designed assuming a material homogeneity that does not exist. While some consideration of latewood proportion is considered in the grading of wood products, the macrostructure assumption of material homogeneity has been accepted practice in the structural design of wood products for at least 75 years. Typical macrostructure assumptions ignore the growth rings where wood is assumed to be a homogeneous, orthotropic continuum.

The annual rings of softwoods are visually obvious and represent cylindrical layers of primarily cellulosic material that possess significantly different properties. The formation of earlywood and latewood is one of the manifestations of weather-based and climate-based events that occur each season. Earlywood is usually defined as the material grown at the beginning of the growing season, with large cells and relatively thin walls, while latewood is the darker colored material typically contained in the last portion of the growth ring. The growth rings are not completely uniform in width and often provide a record of weather and climate-based events that had an impact on growth. These events have impacts on the formation of the tapered cylindrical layer of earlywood in the early part of the growing season followed by the formation of the layer of latewood in the latter part of the growing season. The earlywood and latewood bands represent the mesostructure of wood. The microstructure is made up of individual cells. The cell walls themselves contain small fibers. The ratio of specific gravity of wood from latewood to earlywood growth rings in loblolly pine is typically 2 to 1 or greater. The elastic moduli of latewood are about a factor 2 to 3 greater than for earlywood [48]. The move to managed tree plantations has generally resulted in wider rings with greater proportions of earlywood, expected outcomes from strategies that include thinning, pruning and fertilization. The viscoelastic properties of wood have been associated with those of the lignin within the wood, which exhibits a glass transition temperature near 100°C. The symmetry of wood is orthotropic. Wood has a high degree of anisotropy as a result of the oriented structure of tubules as shown in Figure 7.12.

Bamboo is similar to wood in that it contains parallel tubules but differs in that the stem of bamboo is hollow. Bamboo differs from wood in detailed structure; it is a hierarchical material [49] that contains fiber-like structural features known as bundle sheaths as well as oriented porosity along the stem axis. Bamboo, moreover, has functional gradient properties in which there is a distribution of Young's modulus across the culm (stem) cross-section. Plant fibrous materials including jute, bamboo, sisal, and bamboo contain lignin and cellulose, in which cellulose microfibrils are embedded in a matrix of lignin and hemicellulose.

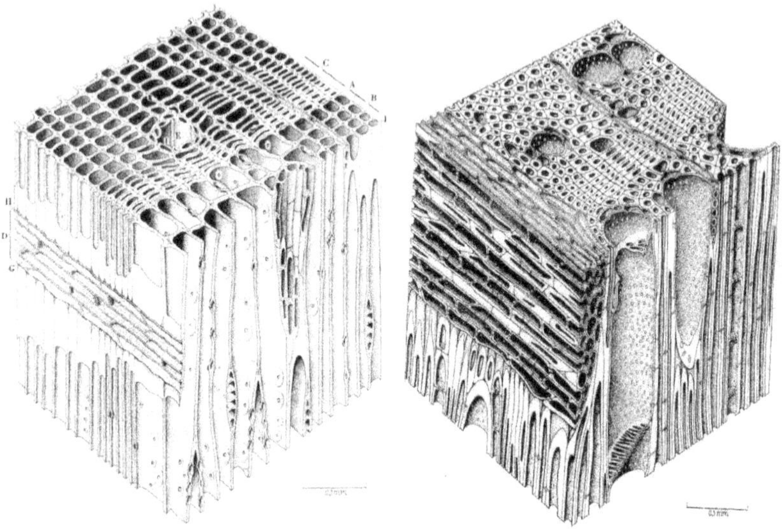

Figure 7.12: Structure of softwood (left) and hardwood (right), [52] with permission. Scale bar, 0.1 mm.

Figure 7.13: Spiral grain in wood.

The grain structure in wood is generally along the axis of the tree trunk. Some trees exhibit spiral grain as illustrated by the Colorado tree trunk shown in Figure 7.13. This wood is structurally chiral. Material chirality is discussed in §3.5. Wood is chiral on a microscopic scale even if the grain structure is aligned with the tree trunk. That chirality has been exploited to make high-performance actuators that twist in response to a change in relative humidity [50]. Trees with spiral grain are less stiff and bend more easily compared with those with straight grain [51]. This may confer an advantage in windy environments. It has also been suggested that spiral grain allows nourishment of the entire tree from roots projecting in a particular direction. This is in contrast to trees with straight grain in which branches are nourished only by roots beneath them. Analysis of the mechanics of trunk deformation does not reveal an obvious advantage to spiral grain.

7.4.2 Wood piezoelectricity

Dry wood is piezoelectric (§4.2), with a sensitivity [53] [54] about 1/20 the d_{11} value (2.3 pC/N) of quartz (§4.2.2). For example, maple had an average d_{14} = -0.08 pC/N; for spruce, the average d_{14} = -0.12 pC/N. The piezoelectric tensor for wood contains only the shear components d_{14} and d_{15}. The piezoelectricity in wood is attributed to piezoelectricity of cellulose, a polar molecule, in the wood. Cellulose has monoclinic symmetry but the crystallites are oriented in wood at random in the plane of symmetry orthogonal to the trunk axis. Piezoelectricity in wood implies chirality of wood. Piezoelectric sensitivity in wood depends on temperature and on the method of drying.

7.5 Summary

Biological materials are composites of biological origin; they have hierarchical structure. The properties of biological materials such as human or animal tissue are of interest to the evolutionary biologist as well as to the biomedical engineer who designs materials to replace or repair damaged tissue. Some biological materials have properties such as ratio of strength to weight, that are competitive with engineering materials. Examples include bone and wood. Designers may draw inspiration from the constitution of these materials.

Bibliography

[1] N. M. Hancox, *Biology of Bone*, Cambridge University Press, Cambridge, (1972).
[2] J. D. Currey, *The Mechanical Adaptations of Bone*, Princeton University Press, Princeton (1984).
[3] R. S. Lakes, Materials with structural hierarchy, Nature, 361, 511-515, (1993).
[4] D. T. Reilly and A. H. Burstein, The elastic and ultimate properties of compact bone tissue, J. Biomechanics , 8, 393-405, (1975).
[5] H. S. Yoon and J. L. Katz Ultrasonic wave propagation in human cortical bone, I. Theoretical considerations for hexagonal symmetry, J. Biomechanics, 9, 407-412, (1976).

[6] S. B. Lang, Elastic coefficients of animal bone, Science, 165 (3890), 287-288, (1969).
[7] S. B. Lang, Ultrasonic method for measuring elastic coefficients of and results of fresh and dried bovine bones, IEEE Trans., Biomedical Engineering, 17, 101-105 (1970).
[8] H. S. Yoon and J. L. Katz, Ultrasonic wave propagation in human cortical bone II. measurements of elastic properties and microhardness, Journal of Biomechanics, 9, 469-464 (1976).
[9] W. C. Van Buskirk, S. C. Cowin, R. N. Ward, Ultrasonic measurement of orthotropic elastic constants of bovine femoral bone, Journal of Biomechanical Engineering, 103, 67-72 (1981).
[10] R. S. Lakes, H. S. Yoon and J. L. Katz, Slow compressional wave propagation in wet human and bovine cortical bone, Science, 220, 513-515, (1983).
[11] J. L. Katz, Hard tissue as a composite material - I. Bounds on the elastic behavior, J. Biomechanics, 4, 455-473, (1971).
[12] P. Frasca, R. A. Harper, and J. L. Katz, Isolation of single osteons and osteon lamellae, Acta Anatomica 95, 122-129, (1976).
[13] R. S. Lakes and S. Saha, Cement line motion in bone, Science, 204, 501-503, (1979).
[14] K. Piekarski, Fracture of bone, J. Appl. Physics 41, 215-223, (1970). Figure reproduced from this article, with the permission of AIP Publishing.
[15] L. J. Gibson and M. F. Ashby, *Cellular Solids*, Pergamon, Oxford, 1988; 2nd Ed., Cambridge University Press, Cambridge (1997).
[16] D. R. Carter, and W. C. Hayes, Bone compressive strength: the influence of density and strain rate, Science 194, 1174-1176, (1976).
[17] J. L. Williams and J. L. Lewis, Properties and an anisotropic model of cancellous bone from the proximal tibial epiphysis, J. Biomechanical Engng. 104, 50-56, (1982).
[18] S. Kaplan, W. C. Hayes, J. L., Stone and G. S. Beaupre, Tensile strength of bovine trabecular bone, J. Biomechanics, 18, 723-727, (1985)
[19] D. R. Carter, G. H. Schwab, and D. M. Spengler, Tensile fracture of cancellous bone, Acta Orthop, Scand. 51, 733-741, (1980).
[20] L. E. Lanyon, W. G. Hampson, A. E. Goodship, and J. S. Shah, Bone deformation recorded in vivo from strain gauges attached to the human tibial shaft, Acta Orthop. Scand., 46, 256-268, (1975).
[21] L. E. Lanyon, Analysis of surface bone strain in the calcaneus of sheep during normal locomotion, J. Biomechanics, 6, 41-69, (1973).
[22] C. T. Rubin, Skeletal strain and the functional significance of bone architecture, Calcif. Tissue Int., 36, S11-S18, (1984).

[23] E. Fukada and I. Yasuda, On the piezoelectric effect of bone, J. Phys. Soc. Japan,12, 1158-1162 , (1957).
[24] C. A. L. Bassett, and R. O. Becker, Generation of electric potentials in bone in response to mechanical stress, Science, 137, 1063-1064, (1962).
[25] J. C. Anderson and C. Eriksson, Piezoelectric properties of dry and wet bone, Nature, 227, 491-492, (1970).
[26] G. Reinish, A. S. Nowick, Piezoelectric properties of bone as functions of moisture content, Nature, 253, 626-627, (1975).
[27] S. B. Lang, Pyroelectric effect in bone and tendon, Nature 212, 704-705, (1966).
[28] J. C. Anderson and C. Eriksson, Electrical properties of wet collagen, Nature, 218, 167-169, (1968).
[29] J. C. Anderson and C. Eriksson, Piezoelectric properties of dry and wet bone, Nature, 227, 491-492, (1970).
[30] G. Galilei, Discorsi E. Dimostrazioni Matematiche intorna a due nuove Scienze, pp. 158-172, 1638, Translated by H. Crew and A. deSalvio, Macmillan, New York, pp. 118-134, (1914).
[31] J. Wolff, *Das Gesetz der Transformation der Knochen*, Hirschwald, Berlin, (1892).
[32] J. A. Spadaro, Electrically stimulated bone growth in animals and man, Clinical Orthopaedics, 122, 325-332, (1977).
[33] D. Brooks, A. H. Burstein, V. H. Frankel, The biomechanics of torsional fractures: the stress concentration effect of a drill hole, J. Bone Joint Surg. 52(3), 507-514 (1970).
[34] J. D. Currey, Stress concentrations in bone, Journal of Cell Science s3-103(61), 111-133, (1962).
[35] J. Kastelik, A. Geliski, E. Baer, The multicomposite structure of tendon, Connective Tissue Res, 6, 11-23, (1978).
[36] R. Vanderby, University of Wisconsin, Madison, WI (2019).
[37] J. Rigby, Effect of cyclic extension on the physical properties of tendon collagen and its possible relation to biological ageing of collagen, Nature 202, 1072, (1964).
[38] R. C. Haut, The mechanical and viscoelastic properties of the anterior cruciate ligament and of ACL fascicles, in *The anterior cruciate ligaments, current and future concepts*, edited by D. W. Jackson, Raven Press, New York, (1993).
[39] B. C. Fleming, B. D. Beynonon, P. A. Renstrom, G. D. Peura, C. E. Nichols, and R. J. Johnson, The strain behavior of the anterior cruciate ligament during bicycling, an in vivo study, Am. J. Sports Medicine, 26, 109-118, (1998).
[40] P. Provenzano, D. Heisey, K. Hayashi, R. S. Lakes and R. Vanderby, Jr., Subfailure damage in ligament: a structural and cellular evaluation, J. Appl. Physiol., 92, 362-371, (2002).

[41] P. Provenzano, K, Hayashi, D. Kunz, M. Markel, R. Vanderby, Jr., Healing of subfailure ligament injury: comparison between immature and mature ligaments in a rat model, J. Orthop. Res. 20, 975-983, (2002).

[42] B. T. Haraldsson, P. Aagaard, M. Krogsgaard, T. Alkjaer, M. Kjaer, and S. P. Magnusson, Region-specific mechanical properties of the human patella tendon, J Appl Physiol. 98: 1006 -1012, (2005).

[43] N. D. Reeves, M. V. Narici, and C. N. Maganaris, Strength training alters the viscoelastic properties of tendons in elderly humans, Muscle Nerve 28, 74-81, (2003).

[44] C. N. Maganaris and J. P. Paul, In vivo human tendon mechanical properties, J. Physiology, 521, 307-313, (1999).

[45] Mechanical properties of wood, D. W. Green, J. E. Winandy, D. E. Kretschmann in *Wood handbook: wood as an engineering material.* Madison, WI: USDA Forest Service, Forest Products Laboratory, (1999).

[46] U. S. Forest Products Laboratory, Madison, WI, lab tech notes, 210, 3, (1952).

[47] R. J. Thomas, Wood: formation and morphology in *Wood structure and composition*, ed. M. Lewin and I. S. Goldstein, Marcel Dekker, New York, (1991).

[48] S. M. Cramer, D. E. Kretschmann, R. S. Lakes and T. W. Schmidt, Earlywood and latewood elastic properties in loblolly pine, Holzforschung, 59, 531-538 (2005).

[49] S. Amada, Hierarchical functionally gradient structures of bamboo, barley, and corn, MRS Bulletin, 20, 35-36, (1995).

[50] N. Plaza, S. L Zelinka, D. S. Stone and J. E Jakes, Plant-based torsional actuator with memory, Smart Materials and Structures, 22(7), 072001 (2013).

[51] S. Leelavanishkul and A. Cherkaev, Why grains in the tree's trunk spiral: mechanical perspective, Structural and Multidisciplinary Optimization, 28, 2-3, 127 - 135 (2004).

[52] U. S. Forest Products Laboratory, courtesy Joseph Jakes (2019).

[53] E. Fukada, Piezoelectricity of wood, Journal of the Physical Society of Japan, 10(2) 149-154, Feb. (1955).

[54] E. Fukada, Piezoelectricity as a fundamental property of wood, Wood Science and Technology, 2(4) 299-307 (1968).

Chapter 8

Size of heterogeneity

8.1 Introduction

Thus far, physical properties of heterogeneous materials have been studied based on the properties of the constituents, the concentration or volume fraction of constituents, and the shape, but not the size, of the heterogeneity. Effective properties have been evaluated using classical theories of elasticity, thermo-elasticity and piezoelectricity in which there is no length scale in the theories. As discussed in §1.1.4 and in Chapter 6, continuum concepts of stress and strain are routinely used for materials that are heterogeneous on a scale that is not negligible in comparison with the scale of experiments or of applications. Composites, foams, and lattices do in fact have a *structural* length scale. In the following, we consider the consequences of heterogeneity size. These consequences include the alteration of stress concentration factors around holes and cracks, size effects in structural rigidity, and coupling between stretching and twisting in chiral materials.

8.2 Stress concentrations

8.2.1 Experiment

There is no length scale in classical elasticity or in classical piezoelectricity or thermo-elasticity. Consequently, there is no predicted sensitivity to gradients of strain. For example the classical theory [1] predicts the stress concentration factor (SCF) for a hole or notch is independent of its size. For anisotropic solids the stress concentration factor differs from that in isotropic solids as discussed in §3.4, but if the material is classically elastic, the stress concentration factor is independent of hole size even in the presence of anisotropy. Experiments are in agreement with theory for metals such as steel and aluminum but not for fibrous composites [2]. Observed stress concentrations around holes depend on hole size (Figure 8.1) in heterogeneous solids. The material is graphite-epoxy laminate with a $[0/\pm 45]_{2s}$ lay-up. In classical elasticity there is no size dependence, even if there is anisotropy. Moreover, the *toughness* of materials has a length scale. For foams (§6.4), the toughness is related to the cell size in the foam.

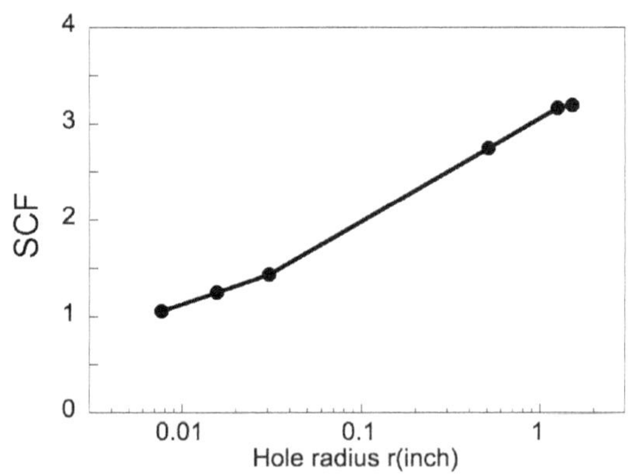

Figure 8.1: Stress concentration factor SCF vs. hole size observed in a fibrous composite, adapted from [2]. In classical elasticity there is no dependence on the hole radius.

8.2.2 Analysis: ad hoc criteria

In the laminated composite community, size effects in stress concentration are well known experimentally [3]. They have been dealt with by using ad-hoc criteria [4]. In this approach, the classical stress distribution around the hole is calculated and one takes the stress value some distance from the hole rather than where the stress is actually maximum (point stress criterion) or an average of stress over a distance (average stress criterion). The distance provides a length parameter to be obtained by curve fitting. Such an approach does not allow one to predict stress concentration factors around a hole of different shape or for stress in a different direction or in other geometries that give rise to a heterogeneous stress field. For example, stress in different directions [5] in experiments revealed a concentration of strain around holes in composites that agreed neither with classical analytical solutions nor with point stress or average stress criteria. Also, such fracture criteria do not allow one to predict size effects in structural stiffness in torsion and bending.

8.2.3 Analysis: generalized elasticity

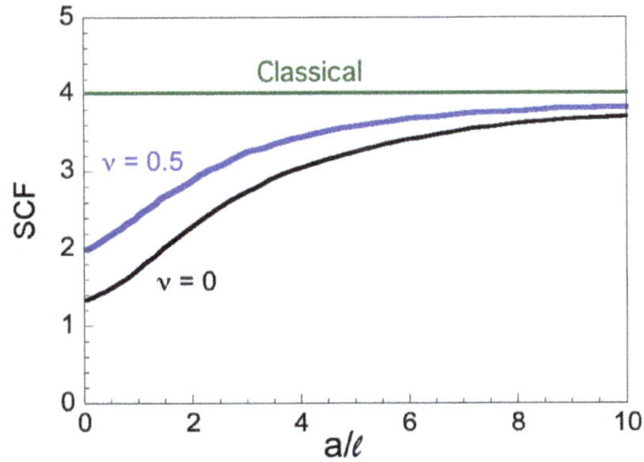

Figure 8.2: Stress concentration factor SCF vs. hole radius a divided by characteristic length ℓ in a field of pure shear predicted by Cosserat elasticity, adapted from [6] for Poisson's ratio 0 and 0.5. In classical elasticity there is no dependence on the hole radius.

Hole size effects in stress concentration [6] are predicted (Figure 8.2) by theories of elasticity that have more freedom than classical elasticity. Some of the freedom is quantified by an additional elastic constant (§8.4), a characteristic length ℓ. Recall that in the context of Poisson's ratio, the amount of freedom to be incorporated in elasticity theory (§3.7.2) was not agreed upon until experiments showed Poisson's ratio to be inconsistent with an early elasticity theory with insufficient freedom. One class of generalized elasticity theories allows a moment per area as well as a force per area (stress). The concept is illustrated in the following example.

8.3 Size effects

8.3.1 Size effects: structural example

To illustrate the effect of constituent size, consider the torsion of a fibrous circular cylinder of radius R. Suppose the concentration of fibers is sufficiently small that $\frac{V_f}{G_f} << \frac{V_m}{G_m}$. For pure shear deformation without any gradients in strain, the predicted shear modulus G_p is slightly greater than the Reuss shear modulus G_R via Halpin - Tsai relations. For small fiber concentration, $G_p \approx G_R \approx G_m(1 - V_f) \approx G_m$. The ratio of torque M to twist angle per length θ is the torsional rigidity which has the classical R^4 dependence on radius R provided the fibers are so small that the moments they carry are negligible.

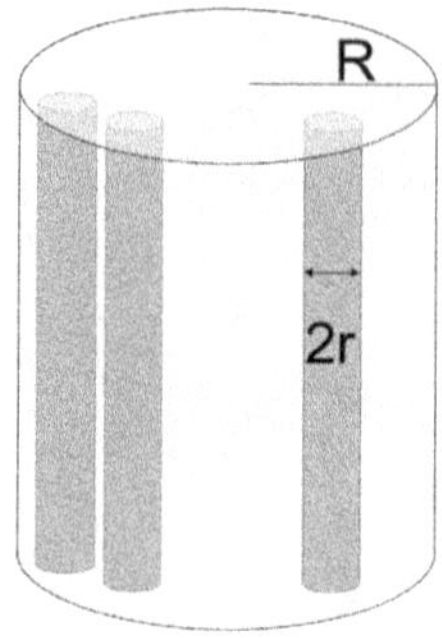

Figure 8.3: Fibrous cylinder of radius R; fibers have a radius r.

$$\frac{M_p}{\theta} = \frac{\pi}{2} G_m R^4. \tag{8.1}$$

Now recognize that in the torsion of the cylinder there is a gradient of rotation that gives rise to the twist of the individual fibers. There are n fibers of radius r and shear modulus G_f so the volume fraction of fibers is $V_f = n(\frac{r}{R})^2$. The moment M_f contributed by the twisting of the fibers gives rise to the following contribution to the rigidity

$$\frac{M_f}{\theta} = n\frac{\pi}{2} G_f r^4. \tag{8.2}$$

Write this to incorporate the cylinder radius R so that the contributions can be expressed in terms of the radius ratio.

$$\frac{M_f}{\theta} = \frac{\pi}{2} G_f V_f r^4 \left(\frac{R}{r}\right)^2 = \frac{\pi}{2} G_f V_f R^4 \left(\frac{r}{R}\right)^2. \tag{8.3}$$

Sum the contributions to rigidity from moments carried by the full cylinder and from torsion of the individual fibers,

$$\frac{M_m + M_f}{\theta} = \frac{\pi}{2} G_f R^4 \left[\frac{G_m}{G_f} + V_f \left(\frac{r}{R}\right)^2\right]. \tag{8.4}$$

or, factoring $\frac{G_m}{G_f}$, the torsion rigidity is

$$\frac{M_m + M_f}{\theta} = \frac{\pi}{2} G_m R^4 \left[1 + V_f \frac{G_f}{G_m} \left(\frac{r}{R}\right)^2\right]. \tag{8.5}$$

The torsional structural rigidity depends on cylinder radius R in a non-classical way. The first term in the [] brackets represents the classical dependence which is dominated by the matrix because $G_p \approx G_R \approx G_m$ under the assumptions made. The second term gives rise to a size effect in which slender cylinders appear to be stiffer than expected from classical elasticity. The nonzero fiber size can be expressed as a structural length parameter $\ell_{str} = r\sqrt{V_f \frac{G_f}{G_m}}$, so

$$\frac{M_m + M_f}{\theta} = \frac{\pi}{2} G_m R^4 [1 + (\ell_{str}/R)^2]. \tag{8.6}$$

8.3.2 Size effects: continuum view

A circular rod of radius R and shear modulus G is now considered as a generalized continuum (§8.4.1) able to support both distributed forces (stress) and distributed moments. The rod exhibits the following structural rigidity, assuming R is sufficiently large compared with ℓ_t or if there is strong coupling between rotation fields in the continuum. The general case is a closed form solution that is rather more complicated [7].

$$\frac{M}{\theta} \approx \frac{\pi}{2} G R^4 [1 + 6(\ell_t/R)^2]. \tag{8.7}$$

The characteristic length ℓ_t is an additional elastic constant in the Cosserat theory of elasticity (§8.4). Compare with the structural length parameter in Equation 8.6. The characteristic length in the generalized continuum view is seen to correspond to the structural length parameter obtained by summing moments in the fibers. This is an elementary approach. More rigorous analyses that extract continuum theory constants from micro-structural features are referred to as homogenization analyses.

In comparing the structural view with the material view, the question arises, when do we consider a solid to be a material. This has a philosophical aspect and has been discussed in §1.1.4.

8.3.3 Size effects: experiment

Milli-scale

Size effects were studied [7] in a composite consisting of round aluminum inclusions in an epoxy matrix in an effort to measure nonclassical effects. The effective shear modulus is independent of specimen size within the stated error estimates as shown in Figure 8.4. This material behaves as a classically elastic material even though it has coarse structure comparable to the size scale of the experiment. Analysis confirms the experiment for the particulate morphology [8].

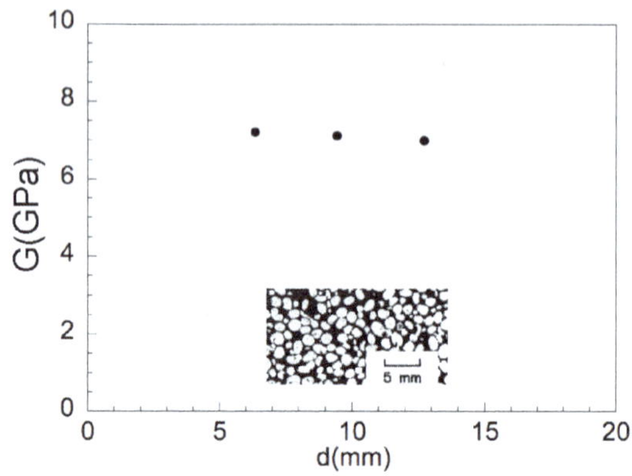

Figure 8.4: Size effects in a particulate composite show classical behavior, adapted from [7]. The inferred shear modulus G is independent of diameter d as is expected via classical elasticity. The inset image is a section of the composite structure.

Size effects in bone are shown in Figure 8.5 [9]. Classical elasticity predicts effective shear modulus independent of specimen size. The experiment shows the apparent modulus increases substantially as the specimen diameter becomes smaller. The Cosserat continuum concept accounts for these observations. From a micro-structural perspective, a contribution to the twisting moment arises from the twist of each individual osteon as a fiber.

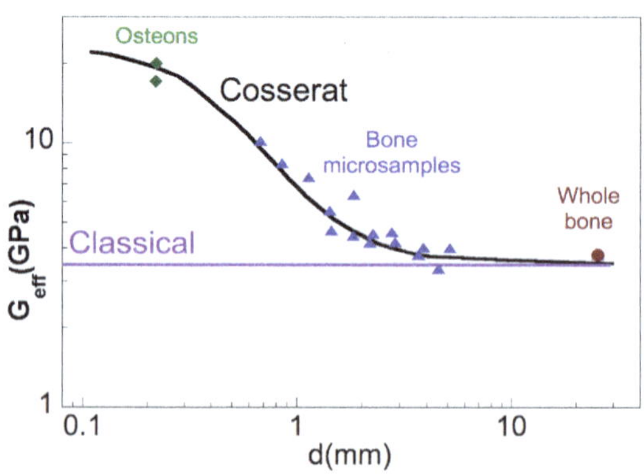

Figure 8.5: Size effects in bone show nonclassical behavior, adapted from [9]. Points represent experimental results. The largest specimen was a whole bone [19]; the smallest specimens were single osteons [20]. G_{eff} is the effective shear modulus. For a classically elastic material, the inferred shear modulus is independent of diameter d as indicated by the horizontal line. The curve is via Cosserat elasticity.

A summary of experimental results for Cosserat elastic constants, mostly extracted from size effects, is discussed as follows and is shown in Table 8.1. Macroscopic scale experiments on solid aluminum [10] showed purely classical behavior hence a characteristic length of zero. Bone [9] [11], foams [12], and designed lattices [13] exhibited characteristic lengths comparable to the structure size. Cosserat concepts have been applied to the dynamics of nano-tubes [14]. Rotational waves of the sort anticipated in Cosserat elasticity were observed in a granular assembly of metal spheres [15]. The granular assembly was predicted to have sensitivity terms α, β, γ in Equation 8.9 equal to zero, hence zero characteristic length. Such a situation is similar to that in composites with hard spheres [8]. Two-dimensional chiral negative Poisson's ratio lattice structures [16], analyzed as Cosserat continua, have characteristic length ℓ comparable to the cell size and large rotation coupling N [17]. Specifically, for rigid circular nodes of spacing S_{node}, connected by slender ligaments of length L_{node}, it was predicted $\ell_t \approx \frac{1}{2} S_{node}$ and $N \approx 0.65$ if $L_{node} << S_{node}$. The stress concentration around holes in these lattices was predicted to be substantially reduced in comparison with a classically elastic solid. Plates with an array of circular holes exhibited size effects [18]; a Cosserat characteristic length was found to be on the order of the hole spacing.

Table 8.1: Summary of elastic constants Young's modulus E, shear modulus G; Cosserat elastic constants are coupling number in torsion N_T and in bending N_B; characteristic length in torsion ℓ_t and in bending ℓ_b. The predominant structure size in aluminum is on the scale of interatomic spacing. Density ρ in g/cc: bone 2.1, dense foam 0.3, open-cell foam 0.03, triangular lattice 0.21.

Material	Structure size (mm)	G (MPa)	E (MPa)	N_T	N_B	ℓ_t (mm)	ℓ_b (mm)
Aluminum[10]	-	-	-	-	-	0	-
Particulate[7]	1.4	7.2×10^3	–	-	-	0	-
Diamond[21]	1.5nm	-	-	-	-	0.2nm	-
Dense foam [22, 23]	0.15	104	300	0.15	0.15	0.3	0.5
Bone[9]	0.25	3.5×10^3	14×10^3	0.6	0.8	0.22	0.44
Foam, open-cell[12]	0.4	0.028	0.081	0.99	0.99	1.6	2.2
Triangular Lattice[13]	9, 10	1.1	3.1	1	0.99	9.4	8.8

The magnitude of observed size effects depends upon ℓ_t, N and upon the range of specimen sizes explored in the experiment or analysis. For dense closed-cell polyurethane foam [22, 23], the ratio of largest to smallest apparent modulus was about a factor of 1.3 associated with the small inferred N. In open-cell foams [12], the size effect exceeds a factor of six; in

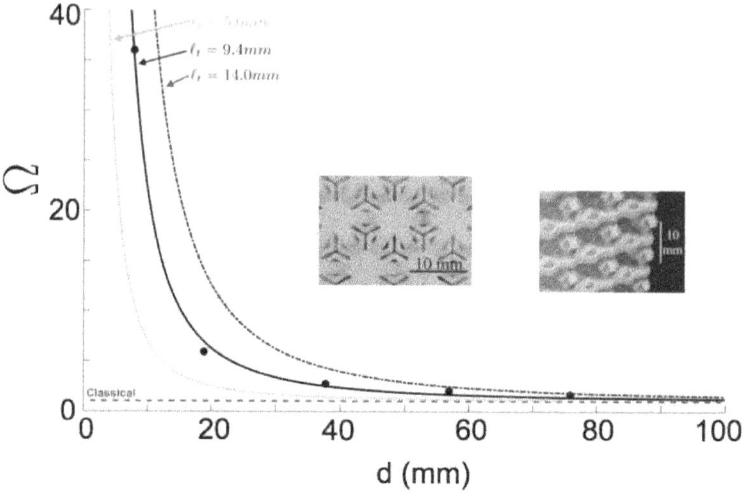

Figure 8.6: Size effects in a 3D printed lattice show strong nonclassical behavior, adapted from [13]. The diameter is d; the ratio of observed structural rigidity to the classical rigidity is Ω. For a classical elastic solid, $\Omega = 1$ as indicated by the horizontal dashed line. The behavior is consistent with Cosserat elasticity. The inset images show sections of the lattice structure.

chiral lattices, a factor of five. For a lattice [13] designed for strong effects, the ratio of observed rigidity to the classical expectation was a factor of about 36 as shown in Figure 8.6.

Hinged structures (§6.6.3, §8.4.5), have been studied as ideal systems. An ideal hinged structure of cubes exhibits anisotropy in Poisson's ratio [24]. Such structures are difficult to realize physically but they may be approximated by using thin flexible ligaments to approximate the hinges. Such a structure is of necessity weak. Some have been made on the milli scale by 3D printing (§6.6.4). The 3D printed solid exhibits modest size effects consistent with Cosserat elasticity because the connecting ribs are not ideal hinges.

Nano-scale

Classical elasticity was predicted to break down in crystalline materials over length scales of 1-10 nm [25]. Chiral cholesteric elastomers [26] were predicted to have a characteristic length on the order of 10 nm.

Fibers at the nano scale are observed to exhibit size effects. Polymer fibers of diameter 1200nm to about 150 nm exhibited a stiffening effect for smaller sizes [27]. Results were interpreted via a strain gradient model; an effective length-scale parameter was found to be 69 nm. Strain gradient theory corresponds to Cosserat elasticity (§8.4.1) for elastic constant $N = 1$. Silicon crystal specimens, by contrast, showed the effective Young's modulus in bending does not change with the specimen width, thickness, and length, hence no size effects [28] for thickness from 200 to 800 nm. A reverse size effect [29] was observed in chromium cantilevers made from thermally evaporated metal. Young's modulus decreased as the cantilevers became thinner from about 80 to 50 nm. This was attributed to a surface effect. Such effects are considered pertinent to the performance of nano-mechanical devices.

Interatomic scale

High frequency waves in diamond exhibited dispersion (dependence of velocity on wavelength or frequency); this was interpreted via Cosserat elasticity with constrained rotation [21]; such wave effects are also understood in the context of structural resonance in the lattice of atoms [30]. The characteristic length calculated from the resulting elastic constants was less than one nanometer. Dispersion of waves was also analyzed [31] based on experimental results in potassium nitrate and interpreted in the context of Cosserat elasticity.

8.4 Generalized continuum elasticity

8.4.1 Cosserat theory

Effects of hole size on stress concentration factors as well as size effects in torsion and bending may be understood in the context of Cosserat elasticity [32] [33], also called micropolar elasticity [34]. Stress concentrations are predicted to be reduced around sufficiently small holes [6] in a Cosserat solid compared with a classical solid.

The Cosserat theory of elasticity is a *continuum* theory that incorporates a local rotation of points as well as the translation assumed in classical elasticity; and a couple stress (a torque per unit area) as well as the force stress (force per unit area). The force stress is referred to simply as stress in classical elasticity, in which there is no other kind of stress. The concept of distributed moments or couple stress was introduced by Voigt in 1887 [39] in the context of the interaction of molecules within single crystals. This was during a time period in which the theory of elasticity itself was under development. Indeed, there was initially disagreement as to the allowable range of Poisson's ratio as discussed in §3.7.2. In 1909 the Cosserat brothers introduced a detailed theory including couple stresses [32]. Some experiments were done to probe the possible applicability of Cosserat elasticity to structural metals such as steel or aluminum but on a macroscopic scale they were found to behave entirely classically. Cosserat elasticity and other generalized elasticity theories then became popular in applied mathematics circles. The subject became of renewed interest following the development of composites, lattices and other materials analyzed thus far.

An isotropic Cosserat solid is describable by six independent elastic constants. In addition to the usual two constants of classical elasticity, there are two characteristic lengths and a coupling constant that governs the magnitude of coupling between local micro-rotations and the macro-rotations associated with gradients of displacement. The final constant is associated with gradients of rotation in different directions and is analogous to Poisson's ratio. These elastic constants can be determined from size effect experiments. The Cosserat equations for stress and couple stress are shown below. This is a *continuum* representation, a generalization of the theory of elasticity. In the corresponding *structural* view, we may envisage increments of force and moment upon ribs in a foam or upon fibers in a fibrous material. The ribs or fibers may rotate with respect to the surrounding material as it is deformed in the structural view, corresponding to the rotation variable in the continuum view. The force per area is stress, σ_{ij} and the moment per area is couple stress, m_{ij}.

$$\sigma_{ij} = 2G\epsilon_{ij} + \lambda\epsilon_{kk}\delta_{ij} + \kappa e_{ijk}(r_k - \phi_k) \tag{8.8}$$

$$m_{ij} = \alpha\phi_{k,k}\delta_{ij} + \beta\phi_{i,j} + \gamma\phi_{j,i} \tag{8.9}$$

Here $\epsilon_{ij} = (u_{i,j} + u_{j,i})/2$ is the strain tensor. The constants λ and G have the same meaning as in classical elasticity; α, β, γ and κ are Cosserat elastic constants. ϕ_i is the local rotation of points (called micro-rotation), and $r_i = (e_{ijk}u_{k,j})/2$ is rotation based on gradients of displacement u.

Classical elasticity is a special case, achieved by allowing α, β, γ, κ to become zero. The classical Lamé elastic constants λ and G then remain and there is no couple stress. The technical constants are as follows. Young's modulus E, shear modulus G, Poisson's ratio ν, characteristic length, torsion ℓ_t, characteristic length, bending ℓ_b, coupling number N, polar ratio Ψ:

$$E = \frac{2G(3\lambda + 2G)}{2\lambda + 2G} \tag{8.10}$$

$$\nu = \frac{\lambda}{2\lambda + 2G} \tag{8.11}$$

$$\ell_t = \sqrt{\frac{\beta + \gamma}{2G}} \tag{8.12}$$

$$\ell_b = \sqrt{\frac{\gamma}{4G}} \tag{8.13}$$

$$N = \sqrt{\frac{\kappa}{2G + \kappa}} \tag{8.14}$$

$$\Psi = \frac{\beta + \gamma}{\alpha + \beta + \gamma}. \tag{8.15}$$

Continuum theories make no reference to structural features; however, they are intended to represent physical solids which always have some form of structure. The couple stresses in Cosserat and microstructure elasticity represent spatial averages of distributed moments per unit area, just as the ordinary (force) stress represents a spatial average of force per unit area. While such moments can occur on the atomic scale or the nano scale, moments may be also transmitted on a much larger scale through fibers in fiber-reinforced materials, or in the cell ribs or walls in cellular solids. The Cosserat characteristic lengths will then be associated with the physical size scales in the microstructure, and be sufficiently large to observe experimentally on the micro-scale or the milli-scale.

8.4.2 Stress and strain fields: effect of microstructure

Generalized continuum theories predict fields of deformation, strain and stress to differ from predictions of classical elasticity. For example, stress concentrations in heterogeneous materials differ from the values predicted classically as discussed in §8.2. Such effects are predicted by Cosserat elasticity (§8.2.3). Stress concentration around a hole is practically important but is difficult to study experimentally because the analyses deal with an infinitely wide plate. Experimentally, the plate should be much larger than the hole size.

Strain concentration in a square cross-section bar in torsion is measurable without the need for a large plate. Classically, the cross-sections warp [1], leading to a peak in strain at the center of the surface and zero strain at the corners. The warp is reduced in a Cosserat solid [35] and the peak strain is reduced. There is more reduction in warp as the characteristic length becomes larger as shown in Figure 8.7. The corresponding strain is redistributed to the corners of the cross-section. Stress can be nonzero at the corners because the stress is allowed to be asymmetric in a material that supports distributed moments to balance the stress.

Predictions of Cosserat elasticity have been demonstrated experimentally using the square cross-section bar configuration. Reduced strain has

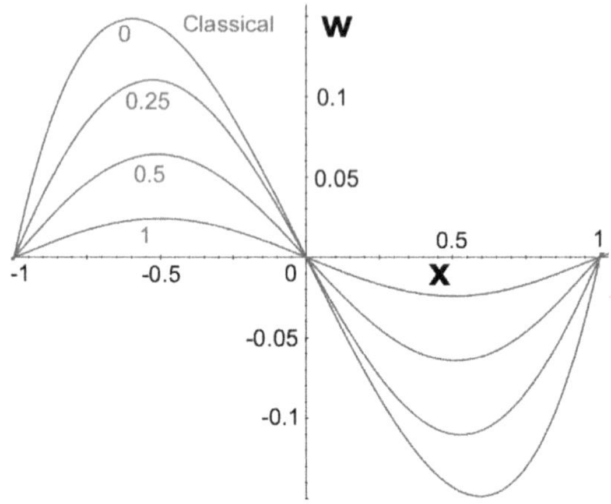

Figure 8.7: Normalized warp $\mathbf{w} = \frac{u_z}{\theta a^2}$ of cross sections of a square section bar in torsion vs. position $\mathbf{x}$ normalized to the half width a, in which u_z is the warp displacement and θ is the twist angle per length. It is assumed that $N = 1$ and that $\ell_b = 0.5\ell_t$. Curves are for $\frac{\ell_t}{a} = 0, 0.25, 0.5, 1$. Classical elasticity corresponds to a characteristic length $\ell_t = 0$.

been observed in bone [36] and reduced warp has been observed in a dense closed-cell foam [37] and in a low density open-cell foam [38].

The effects of microstructure therefore affect the distribution of displacement and strain, reducing the concentration of strain, as predicted by Cosserat elasticity.

8.4.3 Physical causes

Non-classical elasticity arises when the material transmits moments independently of forces. Local moments are easiest to visualize in foams, honeycomb and truss lattices in which moments carried by rib or beam elements are evident in an elementary analysis. Local moments also arise in lattices of atoms that interact via non-central forces; the moments are required to balance the off-center forces. As discussed above, this was recognized by Voigt in 1887 [39].

8.4.4 Homogenization analyses

Various regular structures have been analyzed via homogenization in which the details of the structure are rigorously averaged to obtain a continuum model such as the Cosserat continuum. Characteristic lengths have been extracted from such analyses. For example, composites with stiff spherical inclusions have a predicted characteristic length of zero [8]. Nonclassical effects, if any, will only show up in dynamic wave experiments. As for lattices, the characteristic length for a 2D square lattice of cell size a is predicted to be [42] $\ell = a/2\sqrt{6}$ within couple stress elasticity which corresponds to Cosserat elasticity with $N = 1$. Cosserat elastic constants were determined via homogenization for several lattices [43]. For a 2D square lattice of cell size a, $\ell = a/2\sqrt{6}$ and $N = 1/\sqrt{2}$.

For 2D plates with a regular array of circular holes, [18]; a Cosserat characteristic length was found to be on the order of the hole spacing. Such 2D plates were studied further via finite-element analyses and experimental tests [44]. The values determined for these parameters, particularly the coupling number N, indicate dependence on structure appreciably different from the predictions of lattice theoretical models. The inferred characteristic length was comparable to the size of the holes in the plate.

8.4.5 Hinged structures

Hinged structures (§6.6.3) have been studied in the context of their Poisson's ratio within classical elasticity as discussed above. Structures with ideal hinges do not in general obey the theory of elasticity because they cannot support gradients. A hinged structure of cubes was designed to have a negative Poisson's ratio. The structure is compliant in tension (zero Young's modulus) but is rigid in torsion and also in bending [24]. Classically the bending rigidity should depend only on Young's modulus. That corresponds to a characteristic length that diverges: extreme Cosserat behavior. Similar divergent characteristic length is expected in other hinged structures.

8.4.6 Chirality in elasticity

Chirality has no effect in classical elasticity as discussed in §3.5; by contrast, chirality is well known to be associated with piezoelectricity, pyroelectricity and optical activity. Directional anisotropy and chirality are independent concepts.

Chiral elastic lattices were developed in 2D to exhibit a Poisson's ratio of -1 independent of strain [16]. These 2D lattices were later shown to have the freedom of a Cosserat solid, with a characteristic length ℓ comparable to the cell size and large rotation coupling N [17] as discussed in §8.3.3.

In three dimensions, elastic chirality gives rise to coupling between axial tension or compression and twisting; size effects in stiffness and in Poisson's ratio; and the radial dependence of the Poisson effect [45] via Cosserat analysis.

Stretch-twist coupling effects were observed in thin specimens of bone [40] and in tendon fascicles (fiber bundles) [41]. Chiral cholesteric elastomers [26] were predicted to twist in response to stretching.

Such nonclassical effects such as size effects in rigidity and coupling between squeeze and twist are observed to occur in chiral lattices [46] shown in Figure 6.36 and in other chiral lattices [47]. Stretch-twist coupling occurs in a composite with screw inclusions as shown in Figure 3.4.

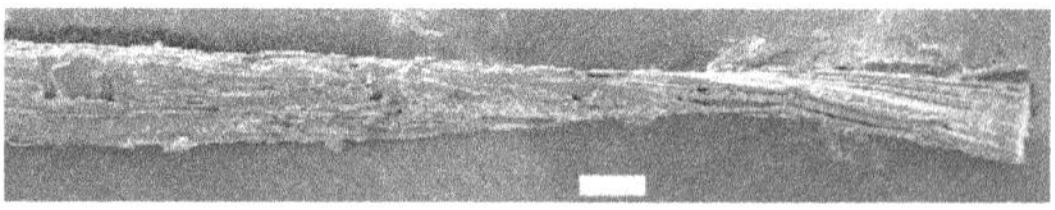

Figure 8.8: Twisting deformation in a wood specimen subjected to a change in humidity [48], with permission. Scale bar, 100μm.

Chiral effects occur in wood specimens. A humidity change causes the twist of wood specimens as shown in Figure 8.8. This chiral effect is the basis of plant-based torsional actuators driven by humidity [49]. The fine scale material chirality is distinct from the structural chirality manifested in some tree trunks that have spiral grain, shown in Figure 7.13.

8.4.7 Other generalized continua

There are several theories of elasticity that provide more freedom than classical elasticity. Cosserat elasticity admits a local rotation of points as well as the translation considered in classical elasticity. One may also provide sensitivity to gradients of dilatation by incorporating a local dilatation variable considered to be associated with void volume change [50]. Void theory predicts size effects in bending but not in torsion. Bending entails a gradient in dilatation but there is no such gradient in torsion. The stress concentration [51] around a hole is larger in void theory than the value predicted by classical elasticity. The stress concentration for a small hole is predicted to be larger than for a large hole. This prediction is the opposite of that of Cosserat elasticity. Thus far, no material has been found that obeys the void theory.

Microstretch elasticity [52] incorporates the sensitivity to dilatation gradient of void elasticity with the sensitivity to rotation gradient of Cosserat elasticity. There are 9 isotropic elastic constants.

The microstructure elasticity theory [53], also called micromorphic, allows a local deformation of points as well as translation as in classical elasticity and rotation as in Cosserat elasticity. The theory allows one to interpret, via a continuum approach, the wave dispersion observed in crystal lattices and in composites. There are 18 isotropic elastic constants.

This theory includes microstretch, Cosserat and classical elasticity as special cases.

Nonlocal elasticity allows stress at a point to depend on strain in a region around that point via an integral over that region [54]. Nonlocal integrals have been approximated as a differential form [55] leading to a sensitivity to strain gradients. Gradient approximations of this type are equivalent to $N = 1$ in Cosserat elasticity but have been called nonlocal.

Experimental methods are available to interpret material behavior in the context of Cosserat elasticity and to extract elastic constants; also void elasticity. The large number of elastic constants in microstructure and microstretch elasticity has thus far precluded experimental methods for their extraction; similarly the nonlocal integral has not been determined from experiments.

8.5 Flexo-electricity: gradient piezoelectricity

Classical piezoelectricity, like classical elasticity, provides no sensitivity to gradients. If the microstructure size is non-negligible in comparison with the size scale of the experiment, sensitivity to gradients can occur in piezoelectricity. Gradient effects in piezoelectricity have long been known [56] [57] and have been interpreted in the context of nonlocal analysis [58]. An electric field applied normal to a polycrystalline layer of barium titanate causes it to bend [56]. Thinner layers bend more than thicker layers. The effect shows a sharp peak near the Curie point at which the material undergoes a phase transformation. Recently such effects have been called flexo-electricity [59] because bending gives rise to electric polarization and an applied electric field causes the bending of a bar made of gradient sensitive material. Bone has long been known to be piezoelectric [60] and flexoelectric [61, 62] and has been recently reviewed [65]. Piezoelectric gradient effects are known in ferroelectric ceramics and in the ferroelectric polymer PVDF. Piezoelectric gradient effects involve a fourth rank coupling tensor in contrast to a tensor of third rank usually associated with piezoelectricity. Therefore such gradient effects do not require chirality in the material. Flexoelectric gradient effect sensitivity is intrinsic to the material. The compression of a cone gives rise to a gradient effect; arrays of cones have been considered in the context of piezoelectric composites [63]. Such composites have been studied in the context of unequal direct and converse effects [64]. Large flexoelectric effects have been observed in polyvinylidene fluoride films [66]. If asymmetric stress is allowed as in Cosserat elasticity, it becomes possible to have piezoelectric sensitivity in an isotropic chiral material [67]. Such a material will exhibit a radial polarization in response to torsional deformation. Ordinarily the lowest symmetry that allows classical piezoelectricity is cubic (§4.2.1).

8.6 Surface and free edge effects

Laminate analysis entails simplifying approximations that ignore details such as local boundary conditions. Actually the stress cannot be uniform or linearly varying within a lamina. Oblique fibers end at the surface of the material. A fiber carries no stress at a free surface, but it acquires stress via transfer through the matrix in the interior of the composite. This gives rise to free edge effects [68] in which local stress distributions may occur near a surface that is nominally free of applied stress. At high applied loads, the material may begin to crack or delaminate prematurely in the vicinity of free surfaces. Free edge effects are pertinent to fatigue performance because fatigue damage leading to failure may initiate at the edges. Stress near free edges as well as stress between laminae, is known to depend upon the stacking sequence of the laminate. Free-edge stress fields tend to decay rapidly with increasing distance from the laminate edge. The region of free-edge stress usually has in-plane dimensions of about one laminate thickness.

In cellular solids such as foam, it is known that surface effects occur in which a layer of incomplete or damaged cells [69] has the effect of reducing the apparent stiffness and strength of the material. These surface effects are more pronounced for smaller specimens. The effect is opposite to the stiffening effect associated with distributed moments.

8.7 Summary

Generalized continuum theories such as Cosserat elasticity, gradient plasticity, and nonlocal elasticity have been recently popular, in part driven by the study of nano-scale phenomena. Nano-scale is not required for such effects to be substantial; it suffices that the largest structure size in the material be non-negligible in comparison with representative size scales associated with gradients in applied fields of stress, electric polarization or other fields. Lattice materials containing rib, plate or surface elements tend to have sufficiently large structure size that such an approach is warranted for macroscopic objects.

Bibliography

[1] S. Timoshenko and J. N. Goodier, *Theory of Elasticity*, McGraw-Hill, New York (1970).

[2] M. E. Waddoups, J. R. Eisenmann, B. E. Kaminski, Macroscopic fracture mechanics of advanced composite materials, J. Composite Materials 5, 446-454, (1971).

[3] B. D. Agarwal and L. J. Broutman, *Analysis and performance of fiber composites*, 2nd Ed. J. Wiley, New York, (1990).

[4] J. M. Whitney and R. J. Nuismer, Stress fracture criteria for laminated composites containing stress concentrations, J. Composite Materials, 8, 253-275 (1974).

[5] L. Toubal, M. Karama, B. Lorrain, Stress concentration in a circular hole in composite plate, Composite Structures 68 31-36 (2005).

[6] R. D. Mindlin, Effect of couple stresses on stress concentrations, Experimental Mechanics, 3, 1-7, (1963).

[7] R. D. Gauthier and W. E. Jahsman, A quest for micropolar elastic constants, J. Applied Mechanics, 42, 369-374, (1975).

[8] D. Bigoni and W. J. Drugan, Analytical derivation of Cosserat moduli via homogenization of heterogeneous elastic materials, J. Appl. Mech. 74, 741-753 (2007).

[9] R. S. Lakes, On the torsional properties of single osteons, J. Biomechanics, 28, 1409-1410, (1995).

[10] J. Schijve, Note on couple stresses, J. Mech. Phys. Solids 14, 113-120, (1966).

[11] J. F. C. Yang, and R. S. Lakes, Experimental study of micropolar and couple stress elasticity in bone in bending, Journal of Biomechanics, 15, 91-98, (1982).

[12] Z. Rueger and R. S. Lakes, Experimental Cosserat elasticity in open cell polymer foam, Philosophical Magazine, 96 (2), 93-111, (2016).

[13] Z. Rueger and R. S. Lakes, Strong Cosserat elasticity in a transversely isotropic polymer lattice, Physical Review Letters, 120, 065501 (2018).

[14] D. Burton and T. Gould, Dynamical model of Cosserat nanotubes, J. Phys. Conf. Ser. 62, 23-33 (2007).

[15] A. Merkel and V. Tournat, Experimental evidence of rotational elastic waves in granular phononic crystals, Phys. Rev. Lett., 107 (22), 225502 (2011).

[16] D. Prall and R. S. Lakes, Properties of a chiral honeycomb with a Poisson's ratio of -1, Int. J. of Mechanical Sciences, 39, 305-314, (1997).

[17] A. Spadoni and M. Ruzzene, Elasto-static micropolar behavior of a chiral auxetic lattice, J. Mechs Physics of Solids, 60, 156-171 (2012).

[18] A. J. Beveridge, M.A. Wheel, D. H. Nash, The micropolar elastic behaviour of model macroscopically heterogeneous materials, International Journal of Solids and Structures 50, 246-255 (2013).

[19] R. Huiskes, J. D. Janssen, and T. J. Slooff, A detailed comparison of experimental and theoretical stress analyses of a human femur, in Mechanical Properties of Bone: Joint ASME-ASCE Applied Mechanics, Fluids Engineering and Bioengineering Conference, Boulder, Colorado, 22-24 June, (1981).

[20] A. Ascenzi, P. Baschieri, and A. Benvenuti, The torsional properties of selected single osteons, J. Biomechanics, 27, 875-884 (1994).

[21] S. Minagawa, K. Arakawa, M. Yamada, Diamond crystals as Cosserat continua with constrained rotation, Physica Status Solidi A 57, 713-718, (1980).

[22] R. S. Lakes, Experimental microelasticity of two porous solids, Int J Solids Structures, 22 55-63, (1986).

[23] Z. Rueger and R. S. Lakes, Experimental study of elastic constants of a dense foam with weak Cosserat coupling, J. Elasticity, 137(1), 101-115, (2019).

[24] C. Andrade, C. S. Ha, R. S. Lakes, Extreme Cosserat elastic cube structure with large magnitude of negative Poisson's ratio, Journal of mechanics of materials and structures (JoMMS), 13(1) 93-101 (2018).

[25] R. Maranganti and P. Sharma, Length scales at which classical elasticity breaks down for various materials. Phys. Rev. Lett. 98 195504 (2007).

[26] M. Warner, E. M. Terentjev, R. B. Meyer, and Y. Mao, Untwisting of a cholesteric elastomer by a mechanical field, Phys. Rev. Lett. 102, 217601 (2009).

[27] L. Sun, R. P. S. Han, J. Wang, C.T. Lim, Modeling the size-dependent elastic properties of polymeric nanofibers. Nanotechnology, 19 455706 (2008).

[28] T. Namazu, Y. Isono, T. Tanaka, Evaluation of size effect on mechanical properties of single crystal silicon by nanoscale bending test using AFM. J. Microelectromech. S., 9, 450-459 (2000).

[29] S. G. Nilsson, X. Borrise, L. Montelius, Size effect on Young's modulus of thin chromium cantilevers, Appl. Phys. Lett., 85, 3555-3557, (2004).

[30] L. Brillouin, *Wave Propagation in Periodic Structures*, Dover, New York, (1953).

[31] A. Askar, Molecular crystals and the polar theories of the continua: Experimental values of material coefficients for KNO3, International Journal of Engineering Science 10(3), 293-300 (1972).

[32] E. Cosserat and F. Cosserat, *Theorie des Corps Deformables*, Hermann et Fils, Paris (1909).

[33] R. D. Mindlin, Stress functions for a Cosserat continuum, Int. J. Solids Structures 1, 265-271 (1965).
[34] A. C. Eringen, Theory of micropolar elasticity. In *Fracture* **1**, 621-729 (edited by H. Liebowitz), Academic Press, New York (1968).
[35] W. J. Drugan and R. S. Lakes, Torsion of a Cosserat elastic bar with square cross section: theory and experiment, Zeitschrift fur angewandte Mathematik und Physik (ZAMP), 69(2), 24 pages (2018).
[36] H. C. Park and R. S. Lakes, Cosserat micromechanics of human bone: strain redistribution by a hydration-sensitive constituent, J. Biomechanics, 19, 385-397 (1986).
[37] W. B. Anderson, R. S. Lakes, and M. C. Smith, Holographic evaluation of warp in the torsion of a bar of cellular solid, Cellular Polymers, 14, 1-13, (1995).
[38] R. S. Lakes, Reduced warp in torsion of reticulated foam due to Cosserat elasticity: experiment, Zeitschrift fuer Angewandte Mathematik und Physik (ZAMP), 67(3), 1-6 (2016).
[39] W. Voigt, Theoretische Studien über die Elasticitätsverhältnisse der Krystalle, Abh. Gess. Wiss. Gottingen, 34, (1887).
[40] R. S. Lakes, Is bone elastically noncentrosymmetric?, Proc. 34th ACEMB. Houston (1981).
[41] K. Buchanan, R. S. Lakes, R. Vanderby, Chiral behavior in rat tail tendon fascicles, Journal of Biomechanics, 64, 206-211 (2017).
[42] C. B. Banks, M. Sokolowski, On certain two-dimensional applications of the couple-stress theory. Int. J. Solids Struct. 4, 15-29 (1968).
[43] A. Spadoni, M. Ruzzene, Elasto-static micropolar behaviour of a chiral auxetic lattice. J. Mech. Phys. Solids 60, 156-171 (2012).
[44] M. McGregor, M. A. Wheel, On the coupling number and characteristic length of micropolar media of differing topology, Proc. Royal Soc. A 470: 20140150 (2014).
[45] R. S. Lakes and R. L. Benedict, Noncentrosymmetry in micropolar elasticity, International Journal of Engineering Science, 29 (10), 1161-1167, (1982).
[46] C. S. Ha, M. E. Plesha, R. S. Lakes, Chiral three-dimensional isotropic lattices with negative Poisson's ratio, Physica Status Solidi B, 253, (7), 1243-1251 (2016).
[47] D. Reasa and R. S. Lakes, Cosserat effects in achiral and chiral cubic lattices, Journal of Applied Mechanics (JAM), 86, 111009-1, 6 pages, (2019).
[48] Image kindly provided by Nayomi Plaza and Tom Kuster at the USDA Forest Service, Forest Products Laboratory, Madison, WI.
[49] N. Plaza, S. L. Zelinka, D. S. Stone and J. E. Jakes, Plant-based torsional actuator with memory, Smart Materials and Structures, 22(7), 072001 (2013).

[50] S. C. Cowin and J. W. Nunziato, Linear elastic materials with voids, J. Elasticity 13 125-147 (1983).
[51] S. C. Cowin, The stresses around a hole in a linear elastic material with voids, Q. J. Mechanics Appl Math 37(3), 441-465 (1984).
[52] A. C. Eringen, Theory of thermo-microstretch elastic solids, Int. J. Engng. Sci., 28(12) 1291-1301, (1990).
[53] R. D. Mindlin, Micro-structure in linear elasticity, Arch. Rational Mech. Analy, 16, 51-78, (1964).
[54] E. Kröner, Elasticity theory of materials with long range cohesive forces, Int. J. Solids and Structures, 3, 731-742, (1967).
[55] A. C. Eringen, On differential equations of nonlocal elasticity and solutions of screw dislocations and surface waves, J. Appl. Phys. 54, 4703-4710 (1983).
[56] E. V. Bursian and O. I. Zaikovskii, Changes in the curvature of a ferroelectric film due to polarization, Sov. Phys. Solid State, 10, 1121-1124, (1968).
[57] N. N. Trunov, Polarization and susceptibility of a ferroelectric sample with a size comparable to the correlation radius, Sov. Phys. Solid State, 17, 1860-1861, (1975).
[58] E. V. Bursian and N. N. Trunov, Nonlocal piezoelectric effect, Sov. Physics Solid State, 16 (4) 760-762, (1974).
[59] I. Naumov, A. M. Bratkovsky, and V. Ranjan, Unusual flexoelectric effect in two-dimensional noncentrosymmetric sp^2-bonded crystals, Phys. Rev. Lett. 102, 217601 (2009).
[60] E. Fukada, and I. Yasuda, On the piezoelectric effect of bone, J. Phys. Soc. Japan, 12, 1158-1162, (1957).
[61] W. S. Williams, Sources of piezoelectricity in tendon and bone, CRC Crit. Rev. in Bioengrg. 2, 95-117, (1974).
[62] W. Williams and L. Breger, Piezoelectricity in tendon and bone, J. Biomech. 8, 407-413, (1975).
[63] J. Fousek, L. E. Cross, D. B. Litvin, Possible piezoelectric composites based on the flexoelectric effect, Materials Letters 39, 287-291, (1999).
[64] J. Y. Fu and L. E. Cross, Separate control of direct and converse piezoelectric effects in flexoelectric piezoelectric composites, Appl. Phys. Lett 91, 162903, (2007).
[65] F. Vasquez-Sancho, A. Abdollahi, D. Damjanovic, and G. Catalan, Flexoelectricity in bones, Adv. Mater., 1705316, (2018).
[66] S. Baskaran, N. Ramachandran, X. He, S. Thiruvannamalai, H. J. Lee, H. Heo, Q. Chen, J. Y. Fu, Giant flexoelectricity in polyvinylidene fluoride films, Physics Letters A 375 2082-2084 (2011).
[67] R. S. Lakes, Third-rank piezoelectricity in isotropic chiral solids, Appl. Phys. Lett., 106, 212905, May (2015).

[68] C. Mittelstedt and W. Becker, Free-edge effects in composite laminates, Appl. Mech. Rev. 60(5), 217-245 (2007).
[69] R. Brezny, and D. J. Green, Characterization of edge effects in cellular materials, J. Materials Sci, 25, 4571-4578 (1990).

Chapter 9

Viscoelastic composites

9.1 Viscoelastic properties: introduction

Viscoelasticity refers to the dependence of the stress-strain relation on time or frequency or rate. Creep refers to an increase of strain with time when stress is held constant. Stress relaxation refers to a decrease of stress with time when strain is held constant. If the stress is prescribed to be sinusoidal in time, the strain will be shifted in phase. Acoustic waves propagated in a viscoelastic material experience attenuation. All these phenomena entail dissipation of mechanical energy. More detail is provided elsewhere [1].

9.2 Viscoelastic functions

9.2.1 Creep

Creep is a progressive deformation of a material under stress that is constant beginning at time zero (Figure 9.1). The history of stress σ as it depends on time t is prescribed to be a step function with magnitude σ_0, beginning at time zero.

The creep compliance is defined as the ratio

$$J(t) = \frac{\epsilon(t)}{\sigma_0}. \tag{9.1}$$

The creep curve may level off to a constant level, may approach a linearly increasing asymptote, or may diverge to failure, depending on the material, the stress level, and the temperature. The specific function of time, which is usually not exponential, depends on the material and temperature. If the material is linearly viscoelastic under the conditions of the test, the creep compliance is independent of stress level. If the load is released at

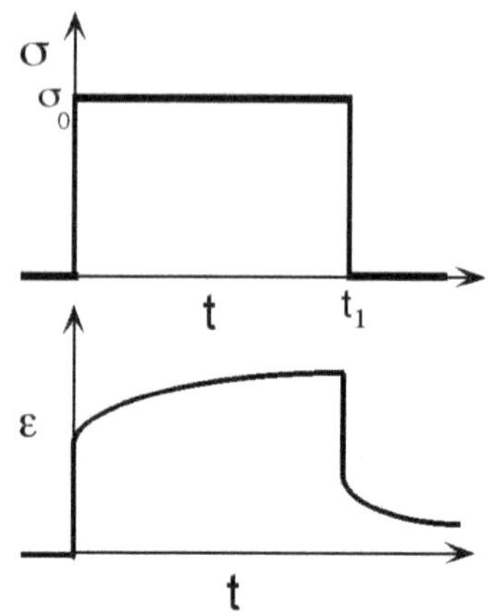

Figure 9.1: Creep: strain $\epsilon(t)$ as a function of time t in response to step stress σ, followed by recovery after the removal of the stress at time t_1.

a later time t_1, the strain will exhibit recovery, or progressive decrease of deformation as shown in Figure 9.1. Strain in recovery may or may not approach zero, depending on the material. The recovery phase can be predicted from the creep if the material is linearly viscoelastic. Creep curves are usually plotted vs. log time.

9.2.2 Relaxation

Stress relaxation is the gradual decrease of stress under strain that is constant beginning at time zero. The history of strain ϵ as it depends on time t is prescribed to be a step function with magnitude ϵ_0, beginning at time zero.

The relaxation modulus is defined as the ratio

$$E(t) = \frac{\sigma(t)}{\epsilon_0}. \tag{9.2}$$

In linearly viscoelastic materials, the relaxation modulus is independent of strain level, so $E(t)$ is a function of time alone. The relaxation modulus may tend to zero or to a nonzero asymptote, depending on the material. The specific function of time, which is usually not exponential, depends on the material and temperature. If the strain is released at a later time t_1, the strain will exhibit recovery as shown in Figure 9.2. The recovery phase can be predicted from the relaxation if the material is linearly viscoelastic. Relaxation curves are usually plotted vs. log time. Time dependence can occur in the bulk modulus, the shear modulus, and in the Poisson's ratio.

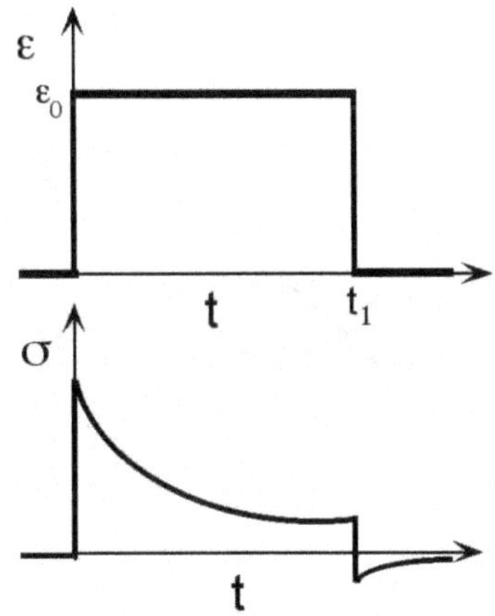

Figure 9.2: Relaxation: stress $\sigma(t)$ as a function of time t in response to step strain ϵ, followed by recovery after the removal of the strain at time t_1.

Creep and relaxation tests reveal viscoelastic response more clearly than stress-strain curves in which time dependence affects the slope, not the magnitude of the curve. It is much easier to discriminate linear from non-linear viscoelastic behavior in creep or relaxation than from a stress strain curve. If the creep compliance is independent of stress in repeated tests, the material is linearly viscoelastic. Similarly if the relaxation modulus is independent of strain in repeated tests, the material is linearly viscoelastic.

9.2.3 Response to sinusoidal input

The response to sinusoidal input is pertinent to applications in which the damping of vibration or the absorption of sound is important. Some experimental methods for characterization are based on sinusoidal input. The stress input is $\sigma(\omega t) = \sigma_0 sin\omega t$, in which t is time, σ_0 is the amplitude, and ω is the angular frequency. Here $\omega = 2\pi f$ with f as the frequency in cycles per second. The strain will be shifted in phase; the phase angle is called δ. The modulus can be represented by a complex quantity: the complex modulus $E^* \equiv E' + iE''$. The primes are conventional notation in this field. The complex modulus depends on frequency. The stress strain curve for a linearly viscoelastic material is an ellipse as shown in Figure 9.3. It can be shown [1] that

$$sin\delta = \frac{A}{B}. \tag{9.3}$$

The loss tangent, tan δ and the absolute value $|E^*|$ of the complex

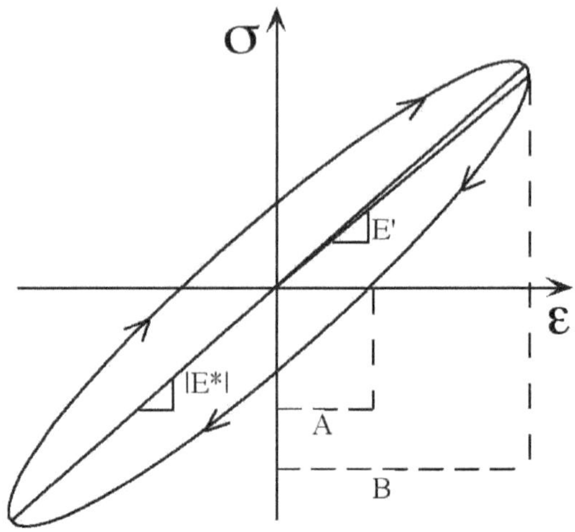

Figure 9.3: Stress $\sigma(t)$ vs. strain ϵ for sinusoidal input.

modulus are typically reported for materials; tan δ is a measure of the damping or dissipation of mechanical energy in the material. If the material experiences structural resonance, then tan δ is proportional to the inverse of the Q or quality factor of the resonance.

It is helpful to know that creep, relaxation and mechanical damping are interrelated. The loss tangent tan δ is related to the slope of the log of the relaxation modulus or creep compliance on a log time scale. If the relaxation function is $E(t) = At^{-n}$, the creep function is $J(t) \propto t^n$ and the loss angle is

$$\delta = \frac{n\pi}{2}. \tag{9.4}$$

An arbitrary creep or relaxation function may be fitted to a power-law over a range of time. The loss tangent calculated is then an approximation.

9.2.4 Viscoelasticity of typical materials

The Young's modulus vs. damping tan δ for some common materials is shown in Figure 9.4. The loss tangent tan δ is never zero; all materials exhibit some viscoelastic response. For a given chemical composition, the modulus is usually well-defined. The damping represented by the loss tangent by contrast may vary over a considerable range. In polymers, the density of cross-links and molecular side groups influence the damping. In crystalline materials such as metals, the damping is influenced by grain boundaries, dislocations, interstitials, vacancies and other defects. Single crystals with few defects of materials such as quartz (silicon dioxide) and sapphire (aluminum oxide or alumina) can exhibit extremely low damping. Rubbery materials (elastomers) can differ substantially in their viscoelastic behavior. Rubbery materials designed for vibration damping and shock isolation can have tan $\delta > 1$ over a narrow range of temperature and frequency.

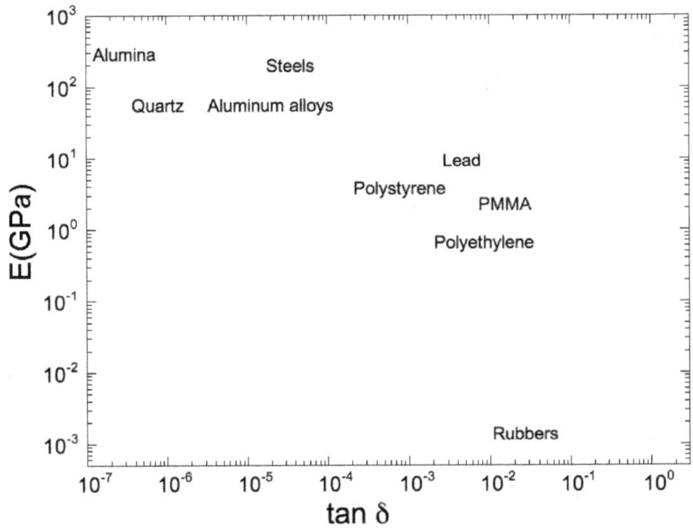

Figure 9.4: Stiffness-damping map. Stiffness, Young's modulus vs. damping tan δ for some common materials.

9.3 Viscoelasticity of composites

The analysis of viscoelastic composites can be done similarly to the corresponding analysis of elastic composites. The Voigt laminate provides a simple illustration.

9.3.1 Viscoelasticity of Voigt laminates

Show that for viscoelastic constituents, $E_c(t) = E_1(t)V_1 + E_2(t)V_2$, for the Voigt composite microstructure, in one dimension, neglecting Poisson effects. Suppose the laminate as a whole is subject to a step function strain in time.

Solution

Each constituent experiences the same strain in the Voigt structure, even if the strain is time dependent. Since it is viscoelastic, the stress-strain relation for phases 1 and 2 is
$\sigma_1(t) = E_1(t)\epsilon_1$, and $\sigma_2(t) = E_2(t)\epsilon_2$
The loads F in terms of the cross-sectional areas A are
$F_1(t) = A_1\sigma_1(t) = E_1(t)\epsilon_1 A_1$, and $F_2(t) = A_2\sigma_2(t) = E_2(t)\epsilon_2 A_2$.
In view of the geometry, the load F_c on the composite block must be the sum of the loads on the constituents.
$F_c(t) = F_1(t) + F_2(t)$,
$F_c(t) = \sigma_c(t)A_c = \sigma_1(t)A_1 + \sigma_2(t)A_2$,
so, dividing by A_c, and recognizing the volume fraction for this geometry as the ratio of cross-sectional areas,
$\sigma_c(t) = \sigma_1(t)\frac{A_1}{A_c} + \sigma_2(t)\frac{A_2}{A_c} = \sigma_1 V_1 + \sigma_2 V_2$.
Finally, divide by the strain, which is the same in both constituents provided they are perfectly bonded.
$E_c(t) = E_1(t)V_1 + E_2(t)V_2$,
as desired.

The method used is one of direct construction. The procedure parallels that used for an elastic composite. By formalizing the similarities in method, one can demonstrate the correspondence principle which allows one to extract the solution to a viscoelastic problem if one knows the solution of the corresponding elastic problem. In the time domain the approach uses Laplace transforms. In the frequency domain associated with sinusoidal input, the correspondence principle is simpler: one replaces an elastic modulus with a complex viscoelastic modulus that depends on frequency [1]. This simplicity enables one to readily explore the effect of different composite microstructures.

9.3.2 Stiffness-damping maps of extremal composites

The behavior of viscoelastic composites is predicted by the correspondence principle [2] [3] [4], by which the relationship between constituent and composite elastic properties is converted to a steady state sine response viscoelastic relation by replacing moduli such as the Young's moduli E by complex moduli $E^*(i\omega)$ or E^*, in which ω is the angular frequency of the harmonic loading. The use of the correspondence principle for the Voigt composite gives for the shear modulus,

$$G_c^* = G_1^* V_1 + G_2^* V_2. \tag{9.5}$$

with $G^* = G' + iG''$ and loss tangent $\tan\,\delta = G''/G'$. Take the ratio of real and imaginary parts. The loss tangent of the composite $\tan\,\delta_c = G''_c/G'_c$ is given by:

$$tan\delta_c = \frac{V_1 tan\delta_1 + V_2 \frac{G'_2}{G'_1} tan\delta_2}{V_1 + \frac{G'_2}{G'_1} V_2}. \tag{9.6}$$

Using the correspondence principle for the Reuss composite, the viscoelastic relation is obtained as

$$\frac{1}{G^*_c} = \frac{V_1}{G^*_1} + \frac{V_2}{G^*_2} \tag{9.7}$$

The corresponding analysis of the loss tangent is simpler in the compliance formulation.

$$tan\delta_c = \frac{V_1 tan\delta_1 + V_2 \frac{J'_2}{J'_1} tan\delta_2}{V_1 + \frac{J'_2}{J'_1} V_2}. \tag{9.8}$$

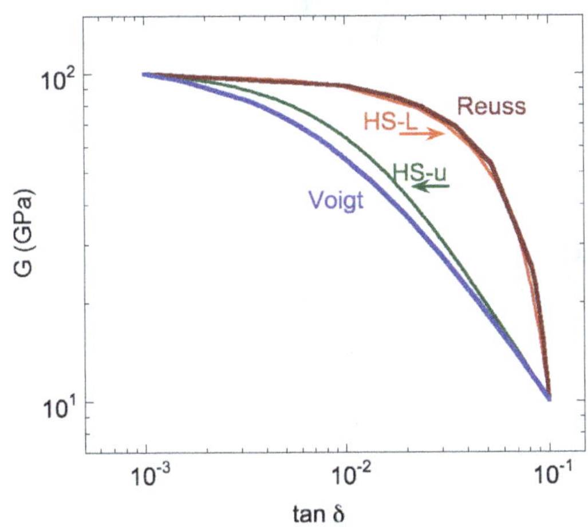

Figure 9.5: Stiffness vs. damping for Voigt, Reuss and Hashin-Shtrikman composites. U refers to the upper bound; L refers to the lower bound; Poisson's ratio is assumed to be 0.3.

The relation between stiffness and damping of the Voigt and Reuss composites is shown in the stiffness-damping map in Figure 9.5 for the full range of volume fractions.

For viscoelastic materials obeying the Hashin-Shtrikman formulae, the correspondence principle is again applied [5]. The complex viscoelastic shear moduli of the composite G_L^* and G_U^* are:

$$G_L^* = G_2^* + \frac{V_1}{\frac{1}{G_1^* - G_2^*} + \frac{6(K_2^* + 2G_2^*)V_2}{5(3K_2^* + 4G_2^*)G_2^*}} \tag{9.9}$$

and

$$G_U^* = G_1^* + \frac{V_2}{\frac{1}{G_2^* - G_1^*} + \frac{6(K_1^* + 2G_1^*)V_1}{5(3K_1^* + 4G_1^*)G_1^*}}. \tag{9.10}$$

The stiffness vs. damping of Hashin-Shtrikman upper and lower composites is shown in comparison with the Voigt and Reuss composites in Figure 9.5. Again, points are obtained for the full range of volume fractions. These formulae are not necessarily bounds for viscoelastic properties.

9.3.3 Stiffness-damping map: inclusion shape

Composites with a dilute concentration of spheres, fibers, and platelets were considered in §2.4. Applying the correspondence principle to the equation for spherical elastic inclusions, Equation 2.20 becomes [5]

$$G_c^* = G_1^* - \frac{15(1 - \nu_1^*)(G_1^* - G_2^*)V_2}{7 - 5\nu_1^* + 2(4 - 5\nu_1^*)\frac{G_2^*}{G_1^*}}. \tag{9.11}$$

for the complex shear modulus of the composite. The loss tangent is computed assuming Poisson's ratio is a real quantity with no phase angle. The behavior of a composite with soft spherical particles resembles that of the Voigt relation in the stiffness-damping map. The behavior of a composite with a dilute concentration of stiff spherical particles resembles that of the Reuss relation in the stiffness-damping map.

For fiber inclusions of random orientation in dilute concentration, applying the correspondence principle,

$$E_c^* = \frac{1}{6}E_2^* V_2 + E_1^* \frac{1 + \frac{1}{4}V_2 + \frac{1}{6}V_2^2}{1 - V_2}. \tag{9.12}$$

For dilute randomly oriented fibers, the stiffening effect of the fibers is $\frac{1}{6}$ of the value obtained in the Voigt composite for which which the fibers are all aligned. In the stiffness-damping plot for viscoelastic composites, the curve for random fibers is similar to the curve for the Voigt solid. Corresponding volume fractions differ.

For platelets of random orientation in dilute concentration, Equation 2.22 becomes, following application of the correspondence principle,

$$E_c \approx E_1^* V_1 + (\frac{1}{2}) E_2^* V_2. \tag{9.13}$$

The curve for platelet inclusions (not shown) in the stiffness-damping map is close to the Voigt and Hashin-Shtrikman upper curve. Either case entails minimal damping for given composite stiffness.

In summary, application of the correspondence principle to the equations for elastic composites enables one to plot stiffness vs. damping in Figure 9.6. A composite containing soft spherical inclusions is similar to

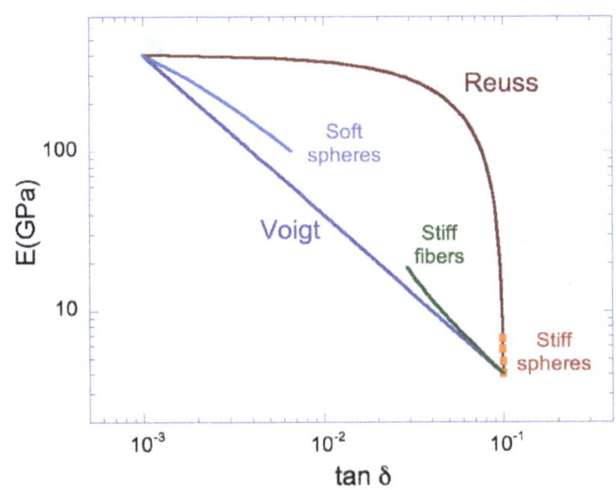

Figure 9.6: Stiffness vs. damping for Voigt and Reuss composites and of composites with a dilute concentration of spheres and fibers.

the Voigt composite in that a small volume fraction of soft, viscoelastic material has a comparatively small effect on the loss tangent [5]. The volume fraction, not shown explicitly in Figure 9.6, will be different. A small concentration of stiff spherical inclusions increased the stiffness while not changing the damping much; this is Reuss-like. Stiff fibers reduce the damping as they increase the stiffness. The effect is similar to that in the Voigt structure. A composite containing soft platelet inclusions has viscoelastic behavior similar to that of the Reuss structure. A small volume fraction of platelet inclusions gives rise to a large increase in loss tangent without reducing the stiffness much. Soft platelets resemble circular cracks in the matrix, so that such a composite would be weakened in comparison with the matrix.

9.3.4 Bounds on viscoelastic properties

Bounds on moduli

Bounds on viscoelastic functions have been obtained [6] from energy principles. The assumption of a non-negative rate of dissipation of energy gives the conclusion $E''\geq 0$. Demonstration of $E'\geq 0$ requires the assumption of both the non-negative stored energy and non-negative rate of dissipation of energy.

Bounds for a composite

Curves for the Voigt and Reuss composites enclose a region in a stiffness loss map. Similarly the curves for the Hashin-Shtrikman composites enclose a region. Such composites are extremal in their elastic properties and they are made using hierarchical structures. The issue of bounds on the viscoelastic behavior is distinct from analyses of bounds on the elastic modulus. Bounds for the real and imaginary parts $E^{'}$ and $E^{''}$ of the complex modulus of composites correspond to the Voigt and Reuss relations [7]. These are not bounds for the stiffness loss map because the damping does not coincide with the maximum in stiffness. Bounds on the viscoelastic behavior in bulk deformation are known for fixed volume fractions of constituents [8] and for arbitrary volume fractions [9]; similarly, bounds for shear [10] [11]. These bounds on a stiffness-loss map are close to or in some cases identical to the curves for the Hashin-Shtrikman composites, which are simpler to express.

9.3.5 Waves in composites

In a classically elastic continuum, plane waves propagate with a speed governed by the elastic properties. Wave speed is independent of frequency provided the wavelength is much larger or much smaller than the specimen dimensions. If the material is viscoelastic, the wave speed increases with frequency and the waves undergo attenuation. A change of wave speed with frequency is called dispersion. The dispersion increases with viscoelastic damping.

The attenuation of waves in one dimension is given by

$$A(z) = A_0 e^{-\alpha z} \tag{9.14}$$

in which the wave amplitude $A(z)$ decreases with the distance of propagation z, A_0 is the initial amplitude, and α is the attenuation per unit length.

The attenuation is related to the damping tan δ by the following, with c as the wave speed.

$$\alpha \approx \frac{\omega}{2c} tan\delta. \tag{9.15}$$

Dispersion and blocking of waves can also arise from resonance in the structural elements of the composite. It has long been known via analysis that laminates exhibit dispersion effects and ranges of frequency in which no waves propagate [12]. Dispersion due to structural causes is called geometric dispersion [13]. The wave speed may decrease with frequency [14]. The frequency at which such effects occur will depend on the size and stiffness of the structural elements. Frequencies in the MHz range are typical for fibrous composites. Extreme dispersion may occur, with large wave attenuation [15]. Cut-off frequencies may occur at which wave propagation is blocked entirely in periodic particulate composites [16] but not in random ones [17].

The intentional use of local resonance in heterogeneous materials has been used to control wave motion as presented in §6.8, including the development of new kinds of lens as well as cloaking.

9.3.6 Negative damping; acoustic amplification

Amplification rather than attenuation of ultrasonic waves can be achieved in piezoelectric semiconductors such as cadmium sulfide [18], [19] in which energy is input via a static electric current. Amplification corresponds to $\tan \delta < 0$, corresponding to a gain tangent rather than a loss tangent. The energy required for amplification is derived from an externally applied DC electric field. The amplification can be as much as a 4% increase in wave amplitude per wavelength. An attenuation of similar magnitude can be achieved depending on the electric field applied. Attenuation is directly related to the damping of $\tan \delta$. The amplification of ultrasonic waves occurs in bismuth [20] subjected to electric and magnetic fields. Negative attenuation or a negative damping, $\tan \delta$, is possible since there is an external energy source. Such materials violate the bounds in §9.3.4 because the energy assumptions do not apply to these materials. In a related vein, a tensed muscle emits low frequency sound over a range of frequency near 25 Hz [21]; the energy source is metabolic in origin. The emission of sound or ultrasound occurs in materials subjected to sufficient static stress. Such acoustic emission is known in metals, steel, ceramics, and wood [22] [23]. The static stress provides energy input to activate the deformation of flaws in the material. The detection of acoustic emission is used in practical methods to evaluate flaws in materials and structures.

9.3.7 Extreme viscoelastic composites: inclusion shape

For the purposes of vibration absorption, it is favorable for a structural element or damping layer to have high stiffness and high damping. It is evident from Figure 9.6 and Figure 9.5 that to simultaneously maximize the modulus and the damping, it is favorable to use stiff particles in a matrix of high damping, or a corresponding hierarchical coated sphere morphology as in Figure 2.8. An attempt to achieve such favorable properties with aluminum matrix containing embedded particles of soft indium [24] resulted in minimal damping at ambient temperature hence poor performance as a material for damping. Such a morphology, with the high damping constituent as the particle is not expected to exhibit the desired properties. By contrast, composites with stiff laminae (tungsten with CdMg) in a Reuss configuration [25] or stiff particulate inclusions (SiC in InSn) [26] exhibited high stiffness and high damping. Similarly, powder of shape memory alloy in indium matrix [27] exhibited high stiffness and high damping.

9.3.8 Extreme viscoelastic composites: stored energy

Bounds on moduli tacitly assume all constituents are in their minimum energy state. That is expressed as a positive definite strain energy. Moduli and spring constants under that assumption are therefore positive. Actually, strain energy may be stored in a lumped or discrete system by buckling. If moduli become negative, damping can become arbitrarily large [28]. Experiments confirm the concept [29]; constrained unit cells containing buckled tubes exhibit damping which diverges as the buckled negative stiffness (§2.5.1) component is tuned to match the positive stiffness component. Strain energy is also stored in composite inclusions in the vicinity of phase transformations. The matrix of a composite containing such inclusions provides a constraint that partially stabilizes the incipient transformation (§2.5.4). Composites of this type exhibited large damping greater than that of either constituent [30]; even a Young's modulus substantially greater than that of diamond [31] combined with high damping, over a small range of temperature. Large damping materials and structures can be useful in reducing noise and vibration.

9.3.9 Viscoelasticity of fibrous composites

Fibrous composites are attractive in applications in which one seeks to maximize stiffness and strength in relation to density (§5.3). Fiber materials such as graphite, boron and glass are not only much stiffer than polymer matrix materials, they also creep much less and have much less damping than polymer matrix [32]. Unidirectional fibrous composites offer the greatest stiffness and strength in the direction of the fibers at the expense of significant anisotropy. The properties depend mostly on those of the fibers so they are called fiber-dominated. Because the modulus of a unidirectional material follows a Voigt construction, the damping is the least in a stiffness-loss map. Minimal damping corresponds to minimal creep following Equation 9.4. The anisotropy of unidirectional fibrous materials can be reduced by making a laminate with fibers in different directions. Such a laminate is less strong and less stiff than a unidirectional composite and will also exhibit more creep and more damping [33]. Laminates in which stiffness, creep and damping are strongly influenced by matrix properties are called matrix-dominated.

In some applications such as vibration absorption, the designer wishes to maximize the viscoelastic response of structural members or added layers. One way to do this is to fabricate composites with a controlled amount of slip between the fibers and the matrix [49]. For steel fibers and a rubber matrix, fibers can be strongly bound via a priming agent or weakly bound by omitting the primer. By varying the surface roughness of the fibers one can control the degree of slip. Large damping and hysteresis is so obtained [49]. Lamination theory applied to graphite epoxy laminates agrees reasonably with the experimental measurement of modulus and damping [47].

Fiber-dominated graphite-epoxy lay-ups such as $[0]_{48}$ and $[0/45/0/-45]_{6s}$ are stiff and exhibit little viscoelastic response and little redistribution of strain due to creep [33]. By contrast, matrix-dominated lay-ups such as $[90]_{48}$ and $[90/-45/90/-45]_{6s}$ are more compliant and exhibit significant creep and strain redistribution.

Graphite-epoxy composites contain fibers with spacings that range from 2 to 7 μm. This length scale is comparable to the size of shear bands, crazes and spherulites in polymer matrix materials; these features influence the viscoelastic response. The behavior of the pure matrix (neat resin) may consequently differ from the same matrix material as a constituent of a composite [50].

9.3.10 Effect of temperature

Viscoelastic effects depend on temperature. In amorphous polymers, the viscoelasticity may follow an Arrhenius-type temperature dependence. If so, one can, by experiments at different temperature, extrapolate behavior to long times or high frequencies that are difficult to probe experimentally. Such a procedure is called time temperature superposition. Master curves refer to plots of viscoelastic properties obtained by shifting curves at different temperature along the log time or log frequency axis. The shift factor refers to the degree of shift. The objective is to obtain a wider range of time or frequency than is ordinarily practicable. In some laminates the shift factor depended on moisture content as well as temperature [48]. One cannot usually invoke time temperature superposition for composites in general because each constituent can have its own temperature dependence.

9.3.11 Poisson's ratio of viscoelastic materials

Poisson's ratio increases with time in polymers because in polymers the shear modulus relaxes much more than the bulk modulus. In composites that contain viscoelastic constituents, each constituent may relax at a different rate. This difference can give rise to complex Poisson's ratios in viscoelastic composites [9]. The viscoelastic Poisson's ratio evaluated in a creep or relaxation test can increase or decrease [35] [36]. Poisson's ratio is a cross property with no energy density associated with it, therefore such effects can occur in passive materials. Poisson's ratio governs the curvature of plates in bending therefore the viscoelastic Poisson's ratio of composite laminates can influence the performance of plate structures. Complex Poisson's ratios have been experimentally determined for some matrix materials and particulate composites [37].

9.3.12 Viscoelasticity of cellular solids

Viscoelasticity in the foam manifests itself in creep under transient tests, and energy dissipation in dynamic studies. Foams exhibit linear viscoelasticity, hence elliptic stress-strain diagrams, under small strain amplitude dynamic loading. A flexible foam or honeycomb will exhibit nonlinear response if deformed sufficiently. If the chemical constitution of the solid phase does not change during the preparation of the cellular solid, then the viscoelastic properties of the cellular solids will follow those of the solid phase.

Foam with air or water in the pores can be analyzed as a fluid-solid composite, as discussed in §4.4. Fluid such as air in the pores of open-cell foams is free to escape as the material is deformed slowly. In a cellular solid with communicating porosity, as in open-cell foams, deformation at sufficiently high frequency gives rise to substantial viscoelastic damping. The cause is the viscous drag of the fluid forced to move through the pores. Flexible open-cell polymer foams can exhibit a maximum tan δ greater than 0.5 [39]. One can extract material constants from viscoelastic experiments upon foam [39], [40] and upon sandstone [41].

In flexible closed-cell foams, the pressure of air or other gas in the pores contributes to the overall stiffness [34]. If a steady load is applied to a closed-cell foam in a creep test, air is lost from the cells due to diffusion through the cell walls. This gives rise to creep [42].

For flexible polymer foams, at sufficiently high frequency (about 1 kHz) the ribs in the foam undergo resonant vibration, giving rise to the attenuation and dispersion of waves [43]. These effects are more pronounced in negative Poisson's ratio foam [44] as a result of the compliance of the ribs.

9.3.13 Viscoelasticity of bone

Compact bone (§7.2) exhibits tan $\delta \approx 0.01$ to 0.02 in the frequency range associated with walking or running. That is smaller than the damping of most polymers but considerably greater than that of structural metals. Damping increases at higher or lower frequency. Bone is too stiff and its damping is too small for it to function as a shock absorber. Soft tissue in the feet helps to ameliorate impacts due to walking or running. Creep over long time periods in bone can be considerable.

9.3.14 Viscoelasticity of tendon and ligament

The viscoelasticity of ligament is apparent to an athlete who performs stretching exercises. The exercise increases the mobility associated with a joint by the time-dependent response of the ligament. During periods of inactivity, viscoelastic strain in the ligament recovers, generating a feeling of tightness. The athlete then repeats the stretching exercise. Tendon and ligament exhibit nonlinear stress strain curves (§7.3). The creep and relaxation response is also nonlinear. The dissipation of energy in the viscoelastic deformation of ligaments is thought to protect the tissue from injury during rapid deformation [45]. Viscoelasticity in living tendon has been observed: it increases with the rate of the slope of the load-deformation curve [46].

9.3.15 Viscoelastic damping of metal matrix composites

Most structural metals exhibit viscoelastic damping of tan δ less than 10^{-3}. Aluminum, for example, exhibits low damping, typically below 10^{-3} for aluminum and below 10^{-5} for some aluminum alloys. Typical inclusion materials in metal matrix composites are refractory ceramics such as silicon carbide or alumina Al_2O_3 which are stiff and have low damping. Silicon carbide (SiC), for example, exhibits $E = 450$ GPa and tan $\delta \approx 10^{-4}$ at 20°C [51]. Most metal matrix composites therefore also have low damping.

Examples of viscoelasticity in particulate metal-matrix composites (§5.5) are as follows. At room temperature, tan $\delta \approx 0.005$ for boron–aluminum, tan $\delta \approx 0.007$ for alumina–aluminum, and tan $\delta \approx 0.004$ for SiC–aluminum [52].

In many applications, particularly at high temperature, it is desirable to reduce creep. In that context, a high volume fraction of fine particle reinforcement substantially reduces creep [53].

If high damping is desired, the use of a high-loss phase such as magnesium alloy can give rise to a high-loss composite [54]. Fullerene inclusions in a magnesium alloy increase damping by about a factor of three.

Damping in metal matrix composites may not follow predictions of elementary composite theory because a constituent may be altered by the formation of the composite. For example, dislocations near the metal-ceramic interface in SiC–Al [55] and in SiC–AlLi [56] have the effect of increasing the damping. Also, thermoelastic damping in composite materials comes from stress-induced heat flow between constituents. Graphite-aluminum composites exhibit tan δ up to 4×10^{-3} due to thermoelastic damping [57]. In this case, damping contains a contribution from a coupled-field interaction between the constituents.

9.4 Summary

Viscoelastic materials have properties that depend on time or frequency or rate. All materials exhibit viscoelastic response. Viscoelastic effects are more pronounced in polymers than in structural metals or in single crystals of refractory materials. Viscoelastic effects in solids tend to increase with temperature. Composite materials allow one to control many aspects of viscoelastic response. Viscoelastic composites allow a high figure of merit for the viscoelastic damping of vibration. It is possible to minimize time-dependent responses such as creep by choice of the composite geometry.

Bibliography

[1] R. S. Lakes, *Viscoelastic Materials*, Cambridge University Press, Cambridge (2009).

[2] R. M. Christensen, Viscoelastic properties of heterogeneous media, J. Mech. Phys. Solids, 17, 23-41, (1969).

[3] Z. Hashin, Complex moduli of viscoelastic composites: I. General theory and application to participate composites, Int. J. Solids, Structures, 6, 539-552, (1970).

[4] R. M. Christensen, *Mechanics of Composite Materials*, John Wiley & Sons, New York, (1979).

[5] C. P. Chen and Lakes, R. S., Analysis of high loss viscoelastic composites, J. Materials Science, 28, 4299-4304, (1993).

[6] R. M. Christensen, Restrictions upon viscoelastic relaxation functions and complex moduli, Trans. Soc. Rheology, 16, 603-614, (1972).

[7] R. Roscoe, Bounds for the real and imaginary parts of the dynamic moduli of composite viscoelastic systems, J. Mech. Phys. Solids, 17, 17-22, (1969).

[8] L. V. Gibiansky and G. W. Milton, On the effective viscoelastic moduli of two phase media: I. Rigorous bounds on the complex bulk modulus, Proc. Royal Soc. London, 440, 163-188, (1993).

[9] L. V. Gibiansky and R. S. Lakes, Bounds on the complex bulk modulus of a two phase viscoelastic composite with arbitrary volume fractions of the components, Mechanics of Materials, 16, 317-331, (1993).

[10] L. V. Gibiansky, G. W. Milton and J. G. Berryman, On the effective viscoelastic moduli of two-phase media: III. Rigorous bounds on the complex shear modulus in two dimensions, Proc. Royal Soc. London, A, 455, 2117-2149, (1999).

[11] L. V. Gibiansky and R. S. Lakes, Bounds on the complex bulk and shear moduli of a two-dimensional two-phase viscoelastic composite, Mechanics of Materials, 25, 79-95, (1997).

[12] S. M. Rytov, Acoustic properties of a thinly laminated medium. Sov. Phys. Acoust., 2, 68-80 (1956).

[13] H. J. Sutherland, On the separation of geometric and viscoelastic dispersion in composite materials, Int. J. Solids, Structures, 11, 233-246, (1975).

[14] H. J. Sutherland, Dispersion of acoustic waves by fiber-reinforced viscoelastic materials, J. Acoustical Soc. of America, 57, 870-875, (1975).

[15] V. K. Kinra and A. Anand, Wave propagation in a random particulate composite at long and short wavelengths, Int. J. Solids, Structures, 18, 367-380, (1982).

[16] V. K. Kinra, E. Ker, An experimental investigation of pass bands and stop bands in two periodic particulate composites, International Journal of Solids and Structures 19(5): 393-410, (1983).

[17] V. K. Kinra, Ultrasonic wave propagation in a random particulate composite, Int. J. Solids, Structures, 16, 301-312, (1980).

[18] A. R. Hutson, J. H. McFee, and D. L. White, Ultrasonic amplification in CdS, Phys. Rev. Lett 7, 237-239, (1961).

[19] D. L. White, Amplification of ultrasonic waves in piezoelectric semiconductors, J. Appl. Phys., 33, 2547-2554, (1962).

[20] A. M. Toxen and S. Tansal, Ultrasonic amplification in bismuth, Phys. Rev. Lett. 10, 481-483, (1963).

[21] G. Oster and J. S. Jaffe, Low frequency sounds from sustained contraction of human skeletal muscle, Biophys. J., 30, 119-127, (1980).

[22] J. Kaiser, Untersuchungen über das Auftreten Gerauschen beim Zugversuch, Arkiv für das Esienhüttenwesen, 24, 43-45, (1953).

[23] C. A. Tatro, A welder's introduction to acoustic emission technology, 1-9, in *Acoustic Emission*, ed. R. W. Nichols, Applied Science Publ. London, (1976).

[24] A. K. Malhotra and D. C. Van Aken, Experimental and theoretical aspects of the internal friction associated with the melting of embedded particles, Acta Metall. Mater., 41, 1337-1346, (1993).

[25] M. Brodt and R. S. Lakes, Composite materials which exhibit high stiffness and high viscoelastic damping, J. Composite Materials, 29, 1823-1833, (1995).

[26] M. Ludwigson, C. C. Swan, and R. S. Lakes, Damping and stiffness of particulate SiC - InSn composite, Journal of Composite Materials, 36, 2245-2254, (2002).

[27] J. San Juan, M. L. Nó, Internal friction in a new kind of metal matrix composites, Materials Science and Engineering A, 442, 429-432, (2006).

[28] R. S. Lakes, Extreme damping in composite materials with a negative stiffness phase, Physical Review Letters, 86, 2897-2900, (2001).

[29] R. S. Lakes, Extreme damping in compliant composites with a negative stiffness phase, Philosophical Magazine Letters, 81, 95-100, (2001).

[30] R. S. Lakes, T. Lee, A. Bersie, and Y. C Wang., Extreme damping in composite materials with negative stiffness inclusions, Nature, 410, 565-567, (2001).

[31] T. Jaglinski, D. S. Stone, D. Kochmann and R. S. Lakes, Materials with viscoelastic stiffness greater than diamond, Science, 315, 620-622, (2007).

[32] J. B. Sturgeon, Creep of fibre reinforced thermosetting resins, in *Creep of Engineering Materials,* ed. C. D. Pomeroy, Mechanical Engineering Publications, Ltd., London, (1978).

[33] M. E. Tuttle and D. L. Graesser, Compression creep of graphite/ epoxy laminates monitored using moiré interferometry, Optics and Lasers in Engineering, 12, 151-171, (1990).

[34] L. J. Gibson and M. F. Ashby, *Cellular Solids*, Pergamon, Oxford, 1988; 2nd Ed., Cambridge, (1997).

[35] R. S. Lakes, The time dependent Poisson's ratio of viscoelastic cellular materials can increase or decrease, Cellular Polymers, 10, 466-469, (1992).

[36] R. S. Lakes and A. Wineman, On Poisson's ratio in linearly viscoelastic solids, J. Elasticity, 85, 45-63, (2006).

[37] A. Agbossou, A. Bergeret, K. Benzarti, and N. Alberola, Modelling of the viscoelastic behaviour of amorphous thermoplastic / glass beads composites based on the evaluation of the complex Poisson's ratio of the polymer matrix, J. Materials Science, 28, 1963-1972, (1993).

[38] Y. M. Wang and G. J. Weng, The influence of inclusion shape on the overall viscoelastic behavior of composites, J. Appl. Mech., 59, 510-518, (1992).
[39] A. N. Gent and K. C. Rusch, Viscoelastic behavior of open cell foams, Rubber Chemistry and Technology, 39, 388-396, (1966).
[40] Y. K. Kim and H. B. Kingsbury, Dynamic characterization of poroelastic materials, Experimental Mechanics 17, 252-258, (1979).
[41] I. Fatt, The Biot-Willis elastic coefficients for a sandstone, J. Applied Mech, 26, 296-297, (1959).
[42] N. J. Mills, Time dependence of the compressive response of polypropylene bead foam, Cellular Polymers, 16, 194-215, (1997).
[43] C. P. Chen and R. S. Lakes, Dynamic wave dispersion and loss properties of conventional and negative Poisson's ratio polymeric cellular materials, Cellular Polymers, 8, 343-359, (1989).
[44] B. Howell, P. Prendergast, and L. Hansen, Acoustic behavior of negative Poisson's ratio materials, DTRC-SME-91/01, David Taylor Research Center, March (1991).
[45] C. Bonifasi-Lista, S. P. Lake, M. S. Small, J. A. Weiss, Viscoelastic properties of the human medial collateral ligament under longitudinal, transverse, and shear loading, J. Orthop. Res, 23, 67-76, (2005).
[46] S. J. Pearson, K. Burgess, G. Onambele, Creep and the in vivo assessment of human patellar tendon mechanical properties, Clinical Biomechanics, 22, 712-717, (2007).
[47] R. D. Adams, Mechanisms of damping in composite materials, J. de Physique, Colloque C9, 44, 29-37, (1983).
[48] D. L. Flaggs and F. W. Crossman, Analysis of the viscoelastic response of composite laminates during hygrothermal exposure, J. Composite Materials, 15, 21-40, (1981).
[49] D. J. Nelson and J. W. Hancock, Interfacial slip and damping in fibre reinforced composites, J. Materials Science 13, 2429-2440, (1978).
[50] S. S. Sternstein, K. Srinavasan, S. Liu and S. Yurgartis, Viscoelastic characterization of neat resins and composites, Polymer Preprints, 25, 201-202, (1984).
[51] A. Wolfenden, P. J. Rynn, and M. Singh, Measurement of elastic and anelastic properties of reaction-formed silicon carbide materials, J. Materials Science, 30, 5502-5507, (1995).
[52] C. R. Wong and S. Holcomb, Damping studies of ceramic reinforced aluminum, in ASTM STP 1169, Mechanics and Mechanisms of Material Damping, DTRC-SME-91/15, David Taylor Research Center, Bethesda, MD, (1991).
[53] D. J. Lloyd, Particle reinforced aluminium and magnesium matrix composites, International Materials Reviews, 39, 1-23, (1994).

[54] H. Watanabe, M. Sugioka, M. Fukusumi, K. Ishikawa, M. Suzuki and T. Shimizu, Mechanical and damping properties of fullerene-dispersed AZ91 magnesium alloy composites processed by a powder metallurgy route, Mater. Trans. 47, (4) 999-1007, (2006).

[55] H. M. Ledbetter and S. K. Datta, Young's modulus and the internal friction of an SiC particle reinforced aluminum composite, Materials Science and Engineering, 67, 25-30, (1984).

[56] I. Gutiérrez-Urrutia, M. L. Nó, J. San Juan, Internal friction behavior in SiC particle reinforced 8090 Al-Li metal matrix composite, Materials Science and Engineering, A 370, 555-559, (2004).

[57] V. K. Kinra, G. G. Wren, S. Rawal and M. Misra, On the influence of ply-angle on damping and modulus of elasticity of a metal-matrix composite, Metall. Trans. 22A, 641-651, (1991).

Appendix A

Appendix

A.1 Solved Problems

A.1.1 Transverse, fibrous

Can the Reuss model be used to obtain the transverse modulus of a unidirectional fibrous composite?

Solution

This will not provide correct results because the geometry is different. In the Reuss laminate the stress is the same in all the layers but in the transverse deformation of a fibrous composite, the stress field around the fibers is complex. The result may be obtained by the theory of elasticity, by finite elements, or by semi-empirical means.

A.1.2 Physical meaning of C_{1111} in applications

Interpret the physical meaning of the modulus C_{1111} in the context of applications and experiments.

Solution

Recall the modulus C_{1111} is related to Young's modulus E by

$C_{1111} = E\frac{1-\nu}{(1+\nu)(1-2\nu)}$.

Since $E = 2G(1+\nu)$ and the Poisson's ratio ν is $\nu = \frac{3K-2G}{6K+2G}$ with G as the shear modulus and K as the bulk modulus, the modulus tensor element may be written

$C_{1111} = B + \frac{4}{3}G$.

The physical meaning of C_{1111} is the stiffness for tension or compression in the x (or 1) direction, when strain in the y and z directions is constrained to be zero. The reason is that for such a constraint the sum in Hooke's law collapses into a single term containing only C_{1111}. The constraint could be applied by a rigid mold, or if the material is compressed in a thin layer between rigid platens. Experiments of this type are done on materials of biological or geological origin. C_{1111} also governs the propagation of longitudinal waves in an extended medium, since the waves undergo a similar constraint on transverse displacement. *Rubbery* materials have Poisson's ratios very close to 1/2, shear moduli on the order of an MPa, and bulk moduli on the order of a GPa. Therefore the constrained modulus C_{1111} is comparable to the bulk modulus and is much larger than the shear or Young's modulus of rubber. If rubber is used in thin layers, it will behave in a stiff manner representative of C_{1111} rather than of Young's modulus E.

A.1.3 Adiabatic and isothermal compliance

Show that in a material with nonzero thermal expansion, the compliance under isothermal conditions differs from the compliance under adiabatic conditions. Find the magnitude of the difference.

Solution

If a simpler approach is desired, the subscripts can be omitted to provide a one dimensional version; they are retained in the following. Consider a unit volume of *elastic* material with strain ϵ_{ij} and entropy S as dependent variables, and stress σ_{ij} and absolute temperature T as independent variables:

$$d\epsilon_{ij} = \frac{\partial \epsilon_{ij}}{\partial \sigma_{kl}}|_T d\sigma_{kl} + \frac{\partial \epsilon_{ij}}{\partial T}|_\sigma dT \tag{A.1}$$

$$dS = \frac{\partial S}{\partial \sigma_{kl}}|_T d\sigma_{kl} + \frac{\partial S}{\partial T}|_\sigma dT \tag{A.2}$$

The elastic compliance is different under deformation which is sufficiently fast that this heat has no time to diffuse (adiabatic condition, $dS = 0$). To calculate the adiabatic compliance, set $dS = 0$ in Equation A.2 and combine Equation A.1, A.2 to eliminate dT.

$$d\epsilon_{ij} = \frac{\partial \epsilon_{ij}}{\partial \sigma_{kl}}|_T d\sigma_{kl} - \frac{[\frac{\partial \epsilon_{ij}}{\partial T}|_\sigma \frac{\partial S}{\partial \sigma_{kl}}|_T]}{\frac{\partial S}{\partial T}|_\sigma} d\sigma_{kl}. \tag{A.3}$$

Now $(\partial \epsilon_{ij}/\partial T)_\sigma = (\partial S/\partial \sigma_{kl})_T$ since these derivatives can be expressed in terms of a thermodynamic potential function by virtue of the first and second laws of thermodynamics. Dividing both sides of Equation A.3 by $d\sigma_{kl}$ to obtain the adiabatic compliance $(\partial \epsilon_{ij}/\partial \sigma_{kl})_S$,

$$\frac{\partial \epsilon_{ij}}{\partial \sigma_{kl}}|_S - \frac{\partial \epsilon_{ij}}{\partial \sigma_{kl}}|_T = -\frac{\partial \epsilon_{ij}}{\partial T}|_\sigma \frac{\partial \epsilon_{kl}}{\partial T}|_\sigma \frac{\partial T}{\partial S}|_\sigma. \tag{A.4}$$

If the material is linear, Equation A.4 becomes

$$J^S_{ijkl} - J^T_{ijkl} = -\alpha_{ij}\alpha_{kl}\frac{T}{C^\sigma} \tag{A.5}$$

Remark: the difference is typically less than one percent for most materials. By contrast, piezoelectric materials can exhibit a large difference in compliance under open-circuit vs. short-circuit conditions.

A.1.4 Foam stiffness vs. density

Show that the stiffness of low density open-cell foam is proportional to the square of the relative density.

Solution

The foam cells have a size proportional to rib length L; ribs are t thick. The stress is $\sigma \propto F/L^2$ with F as load. The deflection in terms of the Young's modulus of the solid is $\delta \propto FL^3/12E_sI$ with $I \propto t^4$. The strain is $\epsilon \propto \delta/L$. The elastic modulus of the foam is $E = \sigma/\epsilon \propto E_s\frac{t^4}{L^4}$. But as discussed in §6.4, $\frac{\rho}{\rho_s} = \frac{t^2}{L^2}$. So the modulus $E \propto E_s(\frac{\rho}{\rho_s})^2$ The constant of proportionality is obtained via a detailed analysis for the particular geometry. For most isotropic foams the constant is close to 1.

A.1.5 Steel foam and solid polymer

What are the true density and the relative density values of steel open-cell foam with the same stiffness (specifically, Young's modulus) as the polymer PMMA (solid, not foam)? Solid steel has a Young's modulus of 200 GPa and a density of 7.8 g/cc; solid PMMA has a Young's modulus 3 GPa and a density of 1.2 g/cc. Is the steel foam more dense or less dense than solid PMMA?

Solution

For foam, $E \propto E_s(\frac{\rho}{\rho_s})^2$ with a constant of proportionality close to 1. Substituting the moduli, the density ratio is 0.122. So $\rho = 0.122\rho_s =$ 0.955 g/cc. This is less dense than 1.2 g/cc, the density of the solid polymer.

A.1.6 Cardboard honeycomb strength

A commercially available cardboard honeycomb sandwich panel is used for packaging, to protect objects from damage during shipping. The cells are regular hexagons with $H = L = 10$ mm; the cell wall thickness is $t = 0.2$ mm. The depth is $B = 50$ mm. The solid phase properties are not known but representative values for paper may be assumed. Suppose $E_s = 5$ GPa and that $\sigma_{y,s} = 0.03\ E_s$. Determine the crushing strength of the honeycomb for compression out-of-plane. If a person strikes the honeycomb hard with an elbow, will the honeycomb crush at a sufficiently small load that will protect the elbow from injury? In the context of compression, ignore the thin face sheets which are 0.25 mm thick.

Solution

The strength, assuming the cell walls fail by elastic buckling via Equation 6.17 is $\frac{\sigma^h_{el,z}}{E_s} = 5.2(\frac{t}{L})^3$.

For plastic buckling, the strength via Equation 6.18 is $\frac{\sigma^h_{pl,z}}{\sigma_{y,s}} = 5.6(\frac{t}{L})^{5/3}$.

The strength based on elastic buckling of the cell walls is 208 kPa. The strength based on plastic buckling is 221 kPa, somewhat higher than the prediction for elastic buckling. The honeycomb will fail at the lower value, that of elastic buckling, so the strength is 208 kPa = 30 psi. Observe that the depth B does not enter the strength calculation but is pertinent to energy absorption which depends on the crushing force and the associated displacement. If the impactor carries too much energy it will bottom out and generate a much larger force. An elbow with a projected area of one square inch will receive a force of 30 pounds during an impact. This should be tolerable. The intended use of the panel is protection of items during shipping, not protecting the elbow.

A.1.7 Poisson's ratio of honeycomb

Suppose force is applied perpendicular to the plane of a honeycomb, so the stress is σ_{zz} with all other stresses zero. Show that the Poisson's ratio of the honeycomb for this direction is the same as that of the solid, ν_{solid}.

Solution

Consider first a solid block of material under load in the z direction. Drill a hole in the z direction. There is no stress on the surface of the hole because the surface is perpendicular to z, so the Poisson effect is unchanged. The same is true if one drills an array of parallel holes. The honeycomb consists of a block with an array of hexagonal holes with parallel axes. Therefore the honeycomb Poisson's ratio for such deformation is the same as that of the solid from which it is made.

A.1.8 Particle inclusion concentration

(a) For identical spherical inclusions in a cubic array, what is the maximum volume fraction?

(b) Suggest an alternate structure that gives rise to a higher volume fraction of particles. Is there an upper limit to the volume fraction?

(c) Give an example of a composite with particulate inclusions. Explain briefly how the heterogeneous structure contributes to the function. What effect would nano-scale particles have?

Solution

(a) Consider a sphere embedded in a cubical unit cell. The composite consists of a cubic array of such unit cells. The volume fraction of inclusions is

$$V_i = \frac{v_{sphere}}{v_{cube}} = \frac{\frac{4}{3}\pi r^3}{(2r)^3} = \frac{\pi}{6} \approx 0.52$$

(b) One can have hexagonal close-packed structure of equal size spheres or a hierarchical structure of spheres of unequal size.

(c) Dental composites consist of polymer matrix with mineral inclusions. The inclusions provide stiffness and wear resistance.

A.1.9 Multiple particle sizes

(a) Let us add more spheres to the above cubical array with their centers at the corners of each cubical unit cell. Calculate how large these spheres can be in terms of r above assuming they are in contact with the original spheres.

(b) What is the volume fraction of spheres in the composite including the new ones added in (b)?

Solution

The large spheres have their centers at the center of a cubical unit cell. The distance from the center to the corner is $r\sqrt{1^2+1^2+1^2}$. Of this distance, a distance r is within the center sphere. So the radius of the corner sphere is $r_{sm} = r(\sqrt{3}-1)$.

The cubical unit cell has 8 corners. Each sphere centered on a corner is shared among 8 other cells. So the volume occupied by the smaller cubes centered on the corners is $\mathbf{V}_{sm} = \frac{8}{8}\frac{4}{3}\pi r^3(\sqrt{3}-1)^3$. The cubical unit cell has side $2r$ so the corresponding volume fraction is $V_{sm} = \frac{\pi}{6}0.392 = 0.205$. The total volume fraction occupied by the spherical inclusions is $0.52 + 0.205 = 0.73$.

A.1.10 Spongy bone modulus

Determine Young's modulus of trabecular bone of density 0.2 g/cm^3. Assume that the tissue behaves as an open-cell foam. Consider that for compact bone, $E = 18$ GPa and the tensile strength is about 140 MPa.

Solution

$\frac{E}{E_{solid}} = (\frac{\rho}{\rho_{solid}})^2$, so $E = 18$ GPa $(0.2/2)^2 = 180$ MPa. So, while compact bone is about ten times stiffer than a glassy polymer or hard plastic, the porosity of trabecular bone lowers the stiffness considerably to about one tenth that of a glassy polymer, more similar to a stiff rubber or a polymer in the leathery regime. The strength does not enter the calculation.

A.1.11 Sneaker sole design

Design a shoe insole to reduce the force of impact during jogging. Consider a human of mass $m = 70$ kg jogging at $v = 3.35$ m/s (about 7.5 mph). Suppose the insole has dimensions $b = 9$ cm by $c = 30$ cm long and is $h = 1$ cm thick, also that the maximum deformation strain is $\epsilon = 0.5$. Specifically find the optimal Young's modulus E. Consider first, rubber, second, polymer foam of relative density 0.1.

Solution

For the rubbery case suppose the deformation is linearly elastic so $F = kx$, $F = Evcx/h = kx$. Suppose the maximum kinetic energy equals the potential energy of deformation so $\frac{1}{2}kx^2 = \frac{1}{2}mv^2$ so $k = mv^2/x^2$. This presumes the mass is fully decelerated during impact. So $E = \frac{mv^2 h}{bcx^2}$, but $\epsilon = x/h$ so $E = \frac{mv^2}{bch\epsilon^2}$.

Substitute the numbers; $E = 12$ MPa, a fairly stiff rubber. If we suppose the runner has a fairly smooth gait, it is not necessary to decelerate the entire body on impact, only the leg. Assuming the leg has an effective mass one tenth that of the whole body, then $E = 1.2$ MPa, a more compliant rubber. If rubber is used, the Poisson effect will be partly restrained by contact with the foot and the shoe. To prevent such a stiffening effect, a ridged structure may be incorporated in the sole to allow transverse expansion during compression.

For foam, the modulus is proportional to the density squared, so the solid polymer of which the foam is made must be $0.1^{-2} = 100$ times stiffer than the corresponding value for rubber, or $E = 120$ MPa for deceleration of the leg. This is in the leathery regime in which considerable viscoelastic behavior may be expected.

The analysis assumes linearity. For foam, deformation will likely occur in the nonlinear regime, which should be considered in a further iteration of design.

Measurements on actual sneaker soles indicate a higher relative density of 0.2 to 0.3, so a more compliant rubbery polymer can be used.

A.1.12 Cubic lattice

Consider a cubic lattice of straight ribs. Density is low enough that stress around joints and the mass of joints can be neglected. Consider ribs that are hollow.

Show that the Young's modulus in a principal direction such as the x direction is given by $\frac{E}{E_s} = C_c[\frac{\rho}{\rho_s}]^n$. Determine n and C_c. The cell size is L, the rib outer dimension is t, and the rib wall thickness is w. Do you really need to know these values to answer the question?

How does this stiffness compare with that of isotropic open-cell foam of equal density? Which material is stiffer? Why?

How many elastic constants are there? Explain why.

Solution

One third of the ribs is oriented in each of the coordinate directions; for low density, the other ribs are considered not to contribute to the stiffness so $C_c = 1/3$. The ribs are aligned and parallel, so a Voigt interpretation is called for, so $n = 1$. So $\frac{E}{E_s} = (1/3)[\frac{\rho}{\rho_s}]$ which is stiffer than foam for which $\frac{E}{E_s} = 1[\frac{\rho}{\rho_s}]^2$. The hollow shape of the ribs has no effect on the modulus. The modulus for compression along a principal direction is governed by the cross-sectional area of the ribs oriented in that direction. The shape does influence the strength. A hollow shape increases the area moment of inertia for given mass, and also increases the buckling strength. Because the relative density is considerably less than 1, the lattice is much stiffer than the foam. The symmetry is cubic, so there are three elastic constants. The principal shear modulus, for example, is governed by the bending of ribs so that the modulus goes as the square of the density.

A.1.13 Lattice with tubular ribs

Consider an octet truss lattice of slender straight ribs of length L. Density is low enough that stress around joints and the mass of joints can be neglected, and that elastic buckling is the failure mechanism in compression.

(a) Show that the compressive strength follows $\frac{\sigma_c^f}{E_s} \propto [\frac{\rho}{\rho_s}]^2$. Do not determine the proportionality constant. The ribs are solid rods of radius r.

(b) Suppose the ribs are made a factor of two thicker and are made tubular (hollow) to maintain the same weight. The length is unchanged. By what factor does the lattice Young's modulus change? Why?

(c) Suppose the ribs are made a factor of two thicker and are made tubular (hollow) to maintain the same weight. The length is unchanged. By what factor does the lattice compressive strength change? Why?

Solution

(a) Recall Equation 6.14, the critical buckling force is $F_{crit} = \frac{n^2\pi^2 E_s I}{L^2}$ in which n is related to the degree of constraint on the ends. The stress for the elastic buckling of ribs goes as $\sigma \propto F_{crit}/L^2 \propto E_s I/L^4$ but for a foam or a 3D lattice, $I \propto t^4$ and $\rho/\rho_s \propto (t/L)^2$. So

$\frac{\sigma}{E_s} \propto (\frac{\rho}{\rho_s})^2$.

(b) The lattice is stretch-dominated so $\frac{E}{E_s} \propto [\frac{\rho}{\rho_s}]$. The weight of the ribs hence the density of the lattice is assumed to be unchanged by making the ribs tubular so E is also unchanged.

(c) The outer radius of a tubular rib is made twice as large as the radius of a solid section rib, $r \to 2r_{solid}$. First find the inner radius so that the weight is unchanged. The length is unchanged so we want the cross-sectional area to be unchanged. $A = \pi r^2_{solid} = \pi r^2 = \pi((2r)^2 - r^2_{inner})$ so $r^2 = 4r^2 - r^2_{inner}$ so $r_{inner} = \sqrt{3}r$. The strength is governed by the cross-sectional moment of inertia. For a solid cross-section rib, $I = \frac{\pi}{4}r^4$; for a tubular rib, $I = \frac{\pi}{4}((2r)^4 - (\sqrt{3}r)^4) = \frac{\pi}{4}7r^4$ so the strength increases by a factor of 7 but the density is unchanged.

A.1.14 Motion from piezoelectric disk

Consider a piezoelectric disk with $d_{33} = 500$ pC/N through the thickness. A 10 volt signal is input. How much motion occurs? How could one obtain more motion?

Solution

Units of pC/N are equivalent to pm/V. So the displacement is $u = Vd$ $= 10$ V 500 pm/V $= 5$ nm. This is not much motion. However if one uses the disk at resonance, the displacement is amplified by the quality factor Q. Recall that $Q = 1/tan\delta$. For a piezoelectric ceramic with $Q = 500$, the displacement becomes 5 nm $\times$ 500 $= 2.5$ μm. This is sufficient to generate detectable ultrasonic waves. Larger motion can be obtained by using a bender configuration in which a thin piezoelectric disk is cemented to a thin metal plate. Expansion of the piezoelectric portion causes bending; for a given strain the displacement is much larger than that in axial deformation. Bender elements are used as acoustic emitters.

A.1.15 Voltage from piezoelectric disk

Consider a piezoelectric disk with $d_{33} = 500$ pC/N through the thickness; the Young's modulus is assumed to be 50 GPa and the dielectric constant $k = 1000$. The disk has a thickness $h = 1$ mm. An impulse compressive stress $\sigma = 50$ MPa is applied. How much voltage is generated?

Solution

The charge q in coulombs is $q = Fd = \sigma Ad$ with A as the cross-sectional area and F as force. Charge is related to voltage V via the capacitance C: $V = q/C$. Recall $C = \epsilon_0 k\frac{A}{t}$ with $\epsilon_0 = 8.85 \times 10^{-12}$ farad/m as the permittivity of space and h as the thickness. So $V = \sigma Ad/(\epsilon_0 k\frac{A}{h})$. So $V = \sigma hd/(\epsilon_0 k) = 2800$ volts. Observe that the stress corresponds to a compressive strain of 0.001; these ceramics can endure higher strains.

Indeed, piezoelectric materials are used in spark generators for stoves and other gas appliances. It is possible to generate more than 7000 volts.

A.1.16 Piezoelectric bender

A piezoelectric bender disc consists of a piezoelectric ceramic disc cemented to a thin circular brass plate. A voltage applied to the ceramic causes a deformation perpendicular to the surface via the sensitivity d_{33}. Why does bending occur? After all, the bending deformation field consists of displacements parallel to the surface not perpendicular to it.

Solution

Piezoelectric ceramics have nonzero d_{33} and d_{31} sensitivity. The sensitivity d_{31} gives rise to a deformation in the x direction via an input electric field in the z direction. That deformation gives rise to the bending.

A.1.17 Laminate of steel and rubber

Consider a laminate of thin layers of steel ($E = 200$ GPa, $\nu = 0.3$) and rubber ($E = 1$ MPa, bulk modulus $K = 2$ GPa) of equal thickness to be used as a vibration isolator. The layers are to be horizontal and the isolator supports a sensitive instrument on a floor that vibrates. What is the effective Young's modulus and effective shear modulus of the laminate? It is intended that the isolator be compliant. Discuss.

Solution

For compression, the rubber is effectively constrained laterally so the modulus to be considered is the C modulus for the rubber. $C = K + \frac{4}{3}G$ but via the isotropic relation $E = 2G(1+\nu)$, G for the rubber which has a Poisson's ratio near 0.5 is 1/3 MPa. So $C = 2$ GPa. The effective compressive modulus is given by a Reuss formula with volume fraction 1/2, $E_{lam} = \frac{200 \times 2 \times 2}{200+2} = 3.96$ GPa. This is much stiffer than the Young's modulus of the rubber because the constraint imposed by the steel layers results in the rubber exhibiting its C modulus, not its Young's modulus.

For shear, the given values for steel imply $G = 77$ GPa. The effective shear modulus of the laminate is given by a Reuss formula, $G_{lam} = \frac{77 \times 10^3 \times 0.33 \times 2}{77 \times 10^3 + 0.33} = 0.66$ MPa.

The isolator is therefore much more compliant in shear than in compression. If compliance in compression is required to isolate vibration in the vertical direction, the rubber could be foamed to reduce the Poisson's ratio, hence C, or it could be prepared in a ribbed configuration to provide room for lateral Poisson expansion.

A.2 Problems and questions

1. Why are airplanes usually made of aluminum rather than steel? After all, the ratio of stiffness to density is similar, as is the ratio of stiffness to weight.

2. Derive the composite stiffness in terms of constituent stiffness values for the Reuss composite microstructure. Neglect Poisson effects.

3. Demonstrate the piezocaloric effect as follows. This effect refers to heating or cooling in a material in response to stress. It is the converse of thermal expansion. Grip a thick rubber band at both ends. Subjectively evaluate the temperature of the band by briefly touching it to a free area of skin such as the bottom of your nose. Suddenly stretch the band and again evaluate the temperature. Does it get warmer or cooler? Keep it stretched for about a minute so it comes into equilibrium with the environment. Evaluate its temperature, suddenly release the tension, and repeat. This time, does it get warmer or cooler? Discuss.

4. If platelets provide the best reinforcement based on theory, why are fibrous laminates more commonly used?

5. Show that $C_{ijkl} = C_{klij}$. Assume the material is describable by a strain energy function.

6. Consider the Voigt composite in three dimensions as a laminate. What is its symmetry? How many elastic constants are there if each phase is elastic? How many have been calculated so far? Can one really neglect Poisson's ratio? What might be done to allow for a nonzero Poisson's ratio in the calculation?

7. Give examples of at least one material of each symmetry.

8. Show that in an orthotropic solid $C_{45} = 0$. Hint: introduce at least one plane of mirror symmetry and write the expanded form of Hooke's law showing all the terms in the sum.

9. Show that if Young's modulus is to be measured by applying stress in the x direction only, Hooke's law in the compliance formulation reduces to one term for Young's modulus as the inverse of a compliance. Hint: expand the sum in Hooke's law in the compliance formulation corresponding to strain in the x direction.

10. Are all chiral materials piezoelectric? Explain.

11. Are any non-chiral materials piezoelectric? Explain.

12. Calculate the thermal expansion of the Voigt and Reuss composites. Neglect Poisson effects.

13. Consider a composite in which one phase is piezoelectric but neither phase is pyroelectric. Can the composite be pyroelectric? Explain.

14. Show that the inversion of all coordinate directions has no effect on the elastic modulus C_{ijkl}.

15. Show that the inversion of all coordinate directions has no effect on the thermal expansion α_{ij}.

16. Show that the inversion of all coordinate directions forces all elements of the piezoelectric sensitivity d_{ijk} to zero. Use the direct inspection method. What do you conclude about the type of crystal classes or textures that can be piezoelectric?

17. Show that the inversion of all coordinate directions forces all elements of the pyroelectric sensitivity p_j to zero.

18. Explain in words why C_{1111} is not the same as Young's modulus.

19. Show that for isotropic solids, $C_{11} = E\frac{1-\nu}{(1+\nu)(1-2\nu)}$. Hint: expand the sum in Hooke's law in the modulus formulation corresponding to stress in the x direction.

20. Why does a unidirectional graphite epoxy fibrous composite have low damping and creep in the fiber direction?

21. If the fiber material is so high in performance, why not make a solid block of that material?

22. Consider a unidirectional graphite epoxy composite. It is loaded axially in tension at a stress one half the the fracture stress in the longitudinal direction. The stress is misaligned from the fiber direction by 15 degrees. Will the composite break? Explain.

23. Determine the best angle for the fibers with respect to the long axis of a cylindrical pressure vessel. Hint: write first the longitudinal and circumferential stress components based on the elementary mechanics of materials. What is the direction of maximum tension?

24. Calculate the reduction in elastic modulus expected from 10% by volume spherical rubber inclusions ($E = 1$ MPa) in polystyrene. Why are such rubber inclusions used in view of the fact they are much more compliant than the matrix?

25. Design a buffer pad to minimize peak impact force from a moving object of mass m moving at speed v. For iteration 1, consider the thickness D to be fixed, consider a linearly elastic material of Young's modulus E. Find the optimal value of the modulus using a one dimensional analysis that neglects Poisson effects. For iteration 2, let the material be a flexible open-cell polymer foam with nonlinearity due to rib elastic buckling the plateau region. For iteration 3, let the material be a polymer foam that is stiff enough that compression causes crushing of the ribs.

26. In the above question, what are some specific examples of each impact absorber?

27. Can isotropic materials be piezoelectric? Consider the fact that the three index permutation symbol is an isotropic tensor of third rank.

28. Show that for the bending of a plate, the figure of merit for structural rigidity per mass is $\frac{E}{\rho^3}$.

29. Determine the thermal expansion of a Reuss composite. Neglect Poisson effects. If one does not neglect Poisson effects, what assumption regarding constituent Poisson's ratios must be made for the result to be exact?

30. Show that the bulk modulus of air at ambient temperature and pressure (100 kPa) is 300 kPa. Hint: consider the ideal gas law. In what context is the bulk modulus of air pertinent?

31. Determine the stress concentration factor for a plate of unidirectional graphite epoxy composite with a circular hole. Assume the plate is subject to tension in the longitudinal direction along the direction of the fibers. Use the data provided and the classical elastic solution of Lekhnitskii. Compare with the value for an isotropic material and discuss.

32. Consider foam for seat cushions.

    (a) If a person of mass 75 kg sits on a seat cushion over an area 1000 cm^2, what is the average pressure in SI units?

    (b) Suppose an isotropic open-cell foam with $E = 100$ kPa, $\nu = \frac{1}{3}$, and $\frac{\rho}{\rho_s} = 0.04$ is used. Will the foam be in the linearly elastic deformation mode, the plateau mode (assumed to be from elastic buckling), or in the densification mode?

(c) What is the stiffness (Young's modulus) of the solid used to make this foam? What kind of solid might it be? Is the suggested foam appropriate? If yes, explain why. If not, suggest a better one.

33. Consider a three-dimensional composite with identical spherical inclusions of radius r. Suppose during processing each inclusion is coated with a layer of matrix $r/10$ thick. The inclusions are then packed in a cubic array, with matrix in the intervening space.

    (a) What is the volume fraction of inclusions?

    (b) How does this composite differ in properties from one in which particles are in contact?

    (c) What is the symmetry of the composite? How many elastic constants are there?

34. Consider a laminate in three dimensions. For load in one direction it behaves as a Voigt material; for load in another direction (such as z) it behaves as a Reuss material.

    (a) Sketch the 3D structure. Show an applied compressive stress along the z axis.

    (b) Determine the composite stiffness for compression along the z axis in terms of the constituent stiffnesses and volume fractions. Make simplifying assumptions if needed; state explicitly any assumptions made.

    (c) If the Poisson's ratio of each constituent is zero, determine *all* of the elastic constants of the composite. Show them as components of the C matrix.

35. A laminate of rank 1 is clearly anisotropic unless the volume fraction V is zero or one. Can a laminate of rank 2 be made isotropic in its dielectric properties?

36. Determine the principal Poisson's ratio for the quasicrystal for which elastic C constants are given in §3.7.3.

Appendix B

Symbols

B.1 Principal symbols and definitions

Principal symbols and definitions are provided below. Some symbols used within different communities have different meanings. For example ν refers to Poisson's ratio and frequency.

Table B.1: Words. (n) refers to neologisms introduced recently.

Word	Meaning
anisotropic	properties depend on direction
auxetic materials (n)	have a negative Poisson's ratio
bio-mimetic materials (n)	design imitates biological structure
bio-inspired materials (n)	design inspired by biological structure
cellular solid	material with void space
creep	strain increases at constant stress
hierarchical	structure within structure
heterogeneous	properties depend on position
honeycomb	periodic two-dimensional cellular solid
isotropic	properties do not depend on direction
lattice	periodic three-dimensional cellular solid
metamaterials (n)	negative or extreme values of physical properties
multifunctional (n)	serve multiple roles in an application
piezoelectric	stress and strain are coupled to electric field
platelet	inclusion with one small dimension
poroelastic	material with a solid phase and fluid in pore space
pyroelectric	electric polarization depends on temperature
relaxation	stress decreases at constant strain
viscoelastic	properties depend on time, frequency, rate

Table B.2: Greek letter symbols

Symbol	Meaning
α	thermal expansion
δ	viscoelastic phase angle
δ_{ij}	Kronecker delta
tan δ	viscoelastic damping
ΔT	temperature change
ϵ	strain
ζ	increment of fluid content
θ	angle
ν	Poisson's ratio
ρ	density
σ	stress
τ	time constant or time variable
ω	angular frequency

Table B.3: Script letter symbols

Symbol	Meaning
$\mathcal{D}$	electric displacement
$\mathcal{E}$	electric field
ℓ_{cell}	cell size
ℓ_t	Cosserat characteristic length

Table B.4: Latin letter symbols

Symbol	Meaning
C_{ijkl}	elastic modulus tensor
C_{ij}	elastic modulus, reduced notation
C	coulomb, unit of electric charge
C	electric capacitance
C_v	heat capacity per volume
d_{ijk}	piezoelectric charge sensitivity tensor
E	elastic Young's modulus
E_s	Young's modulus of solid phase
$E(t)$	viscoelastic relaxation modulus
$E^*(\omega)$	viscoelastic complex modulus
F	force
g_{ijk}	piezoelectric voltage sensitivity tensor
G	elastic shear modulus
h	thickness
H	inverse poroelastic expansion
J	elastic compliance
J_{ijkl}	elastic compliance tensor
k	spring constant
k	dielectric constant
K	elastic bulk modulus
L	length
p_j	pyroelectric coefficient
P	pressure
Q	quality factor
Q	a poroelastic constant
r	radius
R	radius
R	inverse fluid storage coefficient
s	elastic compliance
t	time
V_1	volume fraction, phase 1
V	voltage

Index

CPSIA information can be obtained
at www.ICGtesting.com
Printed in the USA
BVHW061112170720
583885BV00001B/2

9 789811 216367